Twenty Years on Ben Nevis

DEER IN WINTER ON A SPUR OF BEN NEVIS.

[*Regd. Photo. by T. Spence, Chemist, Fort-William.*

CAMBRIDGE LIBRARY COLLECTION

Books of enduring scholarly value

Earth Sciences

In the nineteenth century, geology emerged as a distinct academic discipline. It pointed the way towards the theory of evolution, as scientists including Gideon Mantell, Adam Sedgwick, Charles Lyell and Roderick Murchison began to use the evidence of minerals, rock formations and fossils to demonstrate that the earth was older by millions of years than the conventional, Bible-based wisdom had supposed. They argued convincingly that the climate, flora and fauna of the distant past could be deduced from geological evidence. Volcanic activity, the formation of mountains, and the action of glaciers and rivers, tides and ocean currents also became better understood. This series includes landmark publications by pioneers of the modern earth sciences, who advanced the scientific understanding of our planet and the processes by which it is constantly re-shaped.

Twenty Years on Ben Nevis

Not much is known about the life of William T. Kilgour, apart from the fact that in the late nineteenth century he spent two decades as an irregular member of staff at the meteorological observatory on Ben Nevis. In 1905, a year after the observatory closed due to lack of funds, Kilgour published this account of his experiences, including some of 'the more outstanding incidents inseparable from an existence spent at such an altitude', both as a chronicle of life on the mountain and to encourage the public to support the reopening of the observatory. The text is illustrated with several photographs of the striking natural surroundings as well as images of the meteorologists working and relaxing at the inhospitably located station. The result is an accessible and charming record of scientific life on Britain's highest peak around the turn of the century.

Cambridge University Press has long been a pioneer in the reissuing of out-of-print titles from its own backlist, producing digital reprints of books that are still sought after by scholars and students but could not be reprinted economically using traditional technology. The Cambridge Library Collection extends this activity to a wider range of books which are still of importance to researchers and professionals, either for the source material they contain, or as landmarks in the history of their academic discipline.

Drawing from the world-renowned collections in the Cambridge University Library and other partner libraries, and guided by the advice of experts in each subject area, Cambridge University Press is using state-of-the-art scanning machines in its own Printing House to capture the content of each book selected for inclusion. The files are processed to give a consistently clear, crisp image, and the books finished to the high quality standard for which the Press is recognised around the world. The latest print-on-demand technology ensures that the books will remain available indefinitely, and that orders for single or multiple copies can quickly be supplied.

The Cambridge Library Collection brings back to life books of enduring scholarly value (including out-of-copyright works originally issued by other publishers) across a wide range of disciplines in the humanities and social sciences and in science and technology.

Twenty Years on Ben Nevis

Being a Brief Account of the Life, Work, and Experiences of the Observers at the Highest Meteorological Station in the British Isles

William T. Kilgour

CAMBRIDGE
UNIVERSITY PRESS

University Printing House, Cambridge, CB2 8BS, United Kingdom

Cambridge University Press is part of the University of Cambridge.
It furthers the University's mission by disseminating knowledge in the pursuit of education, learning and research at the highest international levels of excellence.

www.cambridge.org
Information on this title: www.cambridge.org/9781108071987

This edition first published 1905
This digitally printed version 2014

ISBN 978-1-108-07198-7 Paperback

Twenty Years On Ben Nevis

Being a Brief Account of the Life, Work, and Experiences of the Observers at the highest Meteorological Station in the British Isles

By WM. T. KILGOUR

WITH ILLUSTRATIONS

PAISLEY: ALEXANDER GARDNER
Publisher by Appointment to the late Queen Victoria

LONDON:
SIMPKIN, MARSHALL, HAMILTON, KENT & CO., LMD.

PRINTED BY ALEXANDER GARDNER, PAISLEY.

Lofty Ben Nevis, King of Britain's hills,
Wreathed in perpetual snow, thy hoary crest
High heavenward stands, peerless amid the rest;
Unmoved amidst the storm's tempestuous roar
The giant stands, formed by the Mind that wills
And rules the universe by His mysterious power:
Thy ever-changing visage, frowns and smiles
As storm-clouds gather, and the changing light
Of sun and shadow, on thy face awhiles,
Haloes thy brow with matchless tintings o'er;
Or silvery Luna, mistress of the night,
Gleams on the misty billows on thy breast,
Stretching afar a noiseless fleecy sea—
A boundless silent waste, like an Eternity.

G. M.

PREFACE

In launching this unpretentious volume on the ocean of Literature, the Author, while disclaiming for it any literary merit, would fain believe that the life-story of those who, during the past two decades, have laboured in the interest of science on the summit of Ben Nevis, is one calculated to appeal to no limited section of the reading public. Fortified by such a hope, and full of a passionate love for the dear old Ben, he has endeavoured to tell, in simple language, something of the joys and sorrows, the exploits, the vicissitudes, and the reminiscences of the observers; as well as to portray the more outstanding incidents inseparable from an existence spent at such an altitude. He thinks it right, however, to indicate that though not a regular member of the meteorological staff, his was the pleasure, betimes, to take up duty at the lofty station, in addition to having been intimately associated with the institution during the greater part of its existence.

To the late observers, his colleagues, he takes this opportunity of tendering his deep sense of gratitude for their unfailing courtesy in supplying information, and in according him access to records and scientific archives.

Doubly repaid would the Author deem himself if, as a result of the publication of this small work, an indulgent public would agitate for the re-opening of Ben Nevis Observatory, the abandonment of which, in this enlightened age when other countries are devoting so much of their wealth towards the propagation of meteorological research, is, to Britons, alike discreditable and unpatriotic.

FORT-WILLIAM,
May, 1905.

CONTENTS

LIST OF ILLUSTRATIONS

TWENTY YEARS ON BEN NEVIS.

CHAPTER I.

IMPRESSIONS.

TWENTY years on Ben Nevis! To most people, the contemplation of such a protracted isolation from the beaten track of human life would probably result in a series of head-shakes, accompanied by an indrawing of chairs to the ingle-neuk, in order to dispel the shudder which the mere suggestion is calculated to produce. Persons holding such views —and their name is legion—err through ignorance, yet they little wot what untold beauties, what rugged grandeur, and what ineffable splendour of mountain, sea, and sky they know not of. The sublimity of prospect, the variety of phenomena, the rolling mists, and the raging tempests have their own peculiar interest, and none the less enthralling is the study of gales and cloud effects, the torrential rains, the accumulation of snow, and the remarkable range and fluctuation of temperature; but the feature which probably most impresses the uninitiated is the stillness—the awful solitude—which at times prevails amid these fastnesses.

Viewing for the first time this colossal pile which forms the highest peak of the Grampians, a stranger might well soliloquize in the manner of Allan Quatermain when his gaze alighted on Mount

Kenia. Base indeed would be the man who could look upon the mighty snow-wreathed Ben—that white, time-stained sentinel of the years and ages long gone by—and not feel his own utter insignificance; and by whatever name he calls Him, worship God in his heart. Such spectacles are like visions of the soul; they throw wide the window of the chamber of our small selfishness and let in a breath of that air which rushes round the rolling spheres and for a while illumine our darkness with a gleam of the white light which beats upon the throne.

Looking at it from sea-level, one can hardly believe that Ben Nevis has an altitude of 4,406 feet, and that in the British Isles it is impossible for a person to stand on a higher platform. It is really two huge mountains rolled into one, with a lake or tarn nestling between the two shoulders. Vegetation exists up to a height of about 3,000 feet, but beyond that only occasional clumps of verdure are met with. The upper ridges are strewn with boulders of porphyry-granite, and thinking persons making the ascent try to imagine the enormous upheaval which once must have been active on these serrated slopes. The mountain has never apparently been volcanic, and the only feasible explanation of its existence is that here the earth's crust must have been less resisting than in other parts, with the result that the abnormal pressure of internal action found vent, and an evolution followed, forming in

VISITORS WITH PONIES NEGOTIATING THE BRIDLE-PATH.

process of time what every school-boy knows as the highest mountain in the three kingdoms.

To geologists, its crevices and rocks offer much food for thought, and even the smatterer can follow the lines of ice action without much difficulty. A region this, sacred in times past to the foot-fall of the red deer and the hill sheep, its privacy during the last two decades has been not a little disturbed by off-shoots from the madding crowd, whom, after the construction of the bridle-path to the summit, scaled it in thousands during the season.

Many a time have I watched and pitied the toilers trudging along under a broiling sun—sweat-begrimmed and languid, resting, spurting, sucking oranges, drinking water, and, in fact, resorting to all sorts of monkey tricks, as if their objective could be attained a whit sooner by such antics. Some try the exploit lightly shod and suffer accordingly, but it is useless advising them before setting out; *they* have been up hills before and know what's what. Ay! that may be so, but these good folks forget that there are hills and hills!

There is one proper way of doing everything, and the rule is as applicable in the climbing of mountains as in anything else. To make the ascent comfortably and with the least output of exertion, one should wear a pair of strong "tackety" boots, provide himself with a stout walking-stick, and be clad as lightly as possible. A lemon and a few biscuits or sandwiches will not constitute a heavy knapsack, and these rations have frequently proved themselves of the utmost service in an emergency. Unless the pangs of hunger become excessive, it is well to refrain from eating during the climb, and imperatively more so to eschew water. Thirst is the constant companion of the quasi mountaineer, but the craving must be fought against at all hazards, and it will usually be found that a little of the lemon juice will act as a panacea. A resolution should be formed prior to starting, never to rest on the journey up, as the oftener one sits down the harder it becomes to proceed in the rarified atmosphere. Of course with some—ladies especially—this is out of the question, and no rules for their guidance need be laid down. Though prescribed, such maxims would probably be disregarded in any case, so it is as well to steer clear of the possibility of being censured. The fair sex, the broken-winded, and the rheumatic must therefore do their own sweet will on Ben Nevis. A large percentage of seeming incapables reach the top, but in many cases the waste of energy in doing so is bound to have detrimental after-effects.

From Fort-William to the base of the mountain, there is a stretch of about two miles of level road, and the more aristocratic climbers usually cycle or drive over this lap. The bridle-path, which is fully five miles in length, mounts by zig-zag gradients, some of which are as steep as 1 in 5, but in forming the track, advantage was taken of all opportunities to provide an access to the hill-top at the minimum of trouble. With the closing of the Observatory, referred to hereafter, the road will doubtless soon fall into disrepair, but such an eventuality is unlikely to check the peripatetic stream of tourists, who, however, cannot hope in the future to enjoy that hospitality at the highest house in the land so wont to be extended to those who found themselves stranded or lost in the mist in the hours of dark on the hill-top. Their coming can do no harm; indeed, it may have quite an opposite effect, for the more widely it is advertised that Ben Nevis is now without an active Observatory, the sooner will the powers that be recognize the error of their ways in closing such an institution. In order therefore to reach the masses, I couch my invitation in common parlance—"Let 'em all come!"

CHAPTER II.

THE OBSERVATORY.

THE pioneer of meteorological work on Ben Nevis was Mr. Clement L. Wragge, who afterwards held the appointment of Government meteorologist for Queensland at Brisbane. It was he who, under the auspices of the Scottish Meteorological Society, established the first or tentative Observatory on the summit of the Ben, in 1881, climbing the mountain almost daily, and continuing the work during the summer and autumn of 1882. At this time, it must be borne in mind, there was no road to the top, and from what has already been stated, some idea may be formed of the difficulties which beset the path of this intrepid scientist.

A rude stone hut covered with a tarpaulin had been erected on the highest eminence, and here Mr. Wragge after his toilsome journey contentedly brewed for himself a steaming cup of coffee, and, alone with wildest Nature, would partake of his frugal repast, not infrequently to the accompaniment of a blustering Sou'-wester, as he sought what comfort was possible from the somewhat cheerless embers of a tiny fire. His indomitable energy and resource, his untiring zeal for the science of which he was a true disciple, and his determination to secure the establishment of a permanent Observatory

BUILDING THE OBSERVATORY.

on Ben Nevis are worthy of all commendation, and no one grudged him the honours which afterwards fell to his lot.

Some had the temerity to dub him "the inclement rag," and would have it that he was riding his hobby to death, but the enthusiast, who had a soul above such foibles, followed the even tenor of his ways with unconcern. Think of it, you of the rising generation, calling yourselves gentle-folks; you who must, forsooth, board a car rather than walk a few hundred yards: here was a man so entirely wedded to science (he had also a wife who aided and abetted him!) that a daily tramp of over fourteen miles—and that for the greater part on a rugged mountain side—was to him so much good done for his fellows that he gloried in the exploit.

The discussion of the observations taken by Mr. Wragge were most instructive, and went far to justify the expectations which had been formed regarding the part to be played by a permanent station on the hill-top. One of the largest and most important practical problems of meteorology is to ascertain the course which storms follow and the causes which fix that course, so that we may forecast from the phenomena observed not only the certain approach of a storm but the particular course that storm will take. Situated as it is, directly in the track of the south-west storms from the Atlantic which so appreciably affect the weather of Europe, it was felt by the Scottish Meteorological Society that observations taken on Ben Nevis were likely to prove of the utmost value, and an appeal was accordingly made to the public for funds to build and equip an Observatory there. A hearty response was the result, and very soon the subscription list exceeded £5,000, donations varying in amount from £200 to 1d., and the subscribers representing all ranks, from Her late Majesty Queen Victoria downwards.

On securing the necessary capital, steps were at once taken to construct a road to the hill-top in order to get the building material up, and when this had been finished at a cost of about £800, the erection of the Observatory was pushed forward with all speed so that no hitch might occur through the blocking of the path by the winter storms.

Erected of granite blocks, which were plentiful at the site, the original building consisted of one apartment, about thirteen feet square, which had to be utilized alike for living room and office, three small bunks which opened off it, and the relative store rooms, coal cellar, etc., at the other end. Externally the Observatory could not boast of architectural symmetry, but of course the house was erected for a specific purpose and not to please the eye. On account of the rigorous climatic conditions which sometimes obtain at this altitude, it was, moreover, necessary when building the Observatory to provide against all risks in this respect, and the walls near the base were accordingly made from ten to twelve feet thick.

On the 17th of October, 1883, the Observatory was formally opened by Mrs. Cameron-Campbell of Monzie, the then proprietrix of the mountain, and the occasion in many ways was a memorable one. A goodly number braved the elements, and scribes were present from most of the leading journals to record the proceedings, though, unfortunately, they were unable to get their "copy" off from the summit, as telegraphic communication had not then been established. The day was a typical one for Ben Nevis, albeit the ardour of a few Sassenachs must have been considerably damped ere the function terminated. A fitful, gusty wind swirled the fine, powdery snow in great clouds amid the precipices and gullies, and it was comical to watch the figures cut by many in the crowd as they shambled and shivered in about two feet of snow. The ceremony, notwithstanding, passed off most successfully, and the establishment of a permanent Observatory on Ben Nevis had at length become a reality.

During the first winter many hardships were incurred by the recluses at this isolated station, but they bore their lot with equanimity, and laboured conscientiously in the cause to which they had resolved to devote their talents. In planning the Observatory, the architect had failed to guard against certain eventualities, especially in the matter of snow-drift, an omission which, betimes, caused the observers to use stronger language than might have been employed on the plains.

On the approach of winter, huge wreaths quickly accumulated round the house, and as the lower door-way was unprotected, the staff had no option but to adopt navvy tactics to avoid being snowed up. A passage with an upward tendency, on the Esquimaux principle, had to be cut through the snow, and in calm weather this means of communication was easily enough kept clear, but as soon as the wind rose the improvised tunnel was choked in an incredibly short time. Indeed, instances were not awanting where the drift, during the night time, came in faster than one man could shovel it out, and in such cases there was no alternative but to wait till morning, when all hands had to fall in for shovel drill.

This unforeseen obstacle naturally interfered with the continuity of observations, a state of matters which convinced the executive that remedial measures would have to be taken if a thorough record were to be secured. During the following summer, therefore, authority was given for the erection of another room rather larger than the first to serve as laboratory and telegraph office, two additional bedrooms—one of the old ones being converted into a store—a room in which telegrams were received and where visitors might rest, and a tower about thirty feet in height, which latter served the double purpose of carrying a set of anemometers and of providing a convenient exit by way of the flat roof when the lower door-way became snow-bound. The whole

"SHOVEL DRILL."

building, with its outer strong protection of granite, was lined with double wooden walls covered with felt; its windows were likewise all double, and the roof covered with lead, overlaid with snow-boarding.

These somewhat uninteresting details give little indication of the comfort and sense of security afforded by the interior of that domicile, and probably none but those who have lived and laboured there know aught of that sentiment by which the hoary old Ben endeared itself to the heart. What mattered it though the elements held wildest revel without, or that snow and frost athwart the passes vied for the mastery; what though winds roared and storms beat, though lightnings flashed and thunder boomed—the dwellers in that unique resort were full of cheer, and they laughed the foray of Nature to scorn.

Congeniality of task effectually scared ennui, which many erroneously assume figured largely on the observers' diurnal menu; rather let it be known that on a member of the staff descending to the lower regions, he invariably discovered that the constriction of vision, the pollution of atmosphere, and the handiwork of man all combined to pall, after the glorious freedom and immensity of prospect experienced during a sojourn amid the everlasting hills.

Of the daily round at the Observatory, I must not omit to speak, but its phases are numerous, and a separate chapter must be devoted to their delineation.

CHAPTER III.

LIFE AND WORK OF THE OBSERVERS.

THE severity and rigour of the elemental conditions peculiar to Ben Nevis did not permit of self-recording instruments being utilized, a fact which made it necessary to have the various readings personally noted at regular intervals. In order, therefore, to accomplish the prescribed task of taking observations every hour of the night and day, it was expedient that there should be at least two observers always in residence on the summit, so that while one slept, the other should have his weather eye open. No hard and fast rule existed as to the system of watches—this being a matter of arrangement, but the usual regime provided for eight hours duty at night and four during the day.

The man who had kept vigil throughout the silent watches invariably made his last tour of inspection at 5 A.M., and after entering the result of his labours in the daily sheet, there was generally just enough time for him left to bang unmercifully the door of the bedroom in which reposed his colleague, so that the latter might be up and doing for the next scientific scrutiny.

My worst enemy cannot impute to me the stigma of being a dabbler in unparliamentary language—though for this I claim no merit—but ah! friend, that bang: it would have made the proverbial saint swear. Try to imagine the shock of being peremptorily brought back to mundane realities whilst enjoying the most peaceful slumber, stretched at full length on what in truth went by the hackneyed appellation of a bunk, but what to you—inhaling the air of heaven in its pristine purity, and contemplating in vision some charming scene in a far-off Utopia—seemed like a bed of roses. Ay! 'twas hard to leave such a couch, and the task was not simplified by the knowledge that, to obviate breaking the inexorable law of punctuality, boots and clothing had to be donned within a space of five minutes. The observer's garb, naturally, was a rough and ready one; his linen, let us say, equalled x, and to him the practice of the art tonsorial was a negative quantity.

Sauntering into the office with about a minute to spare, the man on day duty, after strapping the note-book over his shoulder and reading the barometer, betook himself outside. In wet weather a rain-guage had to be taken, and the custom was to get inside an oilskin suit. At the thermometer screens he read the temperature, wet and dry, shifted the rain-guage, and then got on to the roof where, on reeording the anemometer reading, he had a series of "eye observations" to commit to his note-book. These comprised the direction and velocity of wind, character and quantity of clouds, visibility, fog, mist, haze, glories, haloes, coronæ, snow, hail, sleet, thunder, lightning, and any other phenomena calling for attention. The sunshine recorder, as well as the maximum and minimum thermometers and rainband were read once daily. During gales, when to venture out would be to court death, most of these observations could be made from the tower by hanging out thermometer boxes, etc.

Returning to the office, the watchman's first duty was to reduce the last readings taken, any spare interval thereafter being devoted towards entering up, checking, and tabulating daily or monthly sheets. His mate of the night-shift was by this time probably in the land of dreams, for insomnia was unknown on Ben Nevis; but instead of concocting schemes of revenge, he, who had himself been so rudely awakened, sought refuge in a cup of coffee. This he made for himself in the kitchen, where the fire was never allowed to go out, but he had to partake of his snack judiciously—that is, without making any din—because the cook slept off this apartment, and to arouse this worthy ere the claims of Morpheus had been fully satisfied, was to spoil catering for the day.

Nine o'clock was the breakfast hour, and the Chef (capital C please, Mr. Printer) generally condescended to be called thirty minutes prior to then, maintaining that he could tidy up, set the table, and turn out an appetible cutlet well within half an hour. In this he did not often fail, and altogether matters culinary were dispensed alike lavishly and

TAKING READINGS AT HALF-WAY STATION.

satisfactorily on the Ben. When breakfast was over, the observer on duty and the cook betook themselves in fine weather to the flat roof for a smoke, and it is difficult to convey in mere words the unalloyed pleasure which accompanied those periods of relaxation.

Lounging in utter abandon in the glorious sunshine, and fanned by soft ozone-charged zephyrs from the Atlantic, their's was happiness personified, and conversation somehow seemed a desecration—enough that the vision should drink in that intoxicating panoramic vista of mountains, lakes, and seas; sufficient that for the time at least there was a fulness of that peace which the heart of man is continually yearning for. But at length the spell was broken; a dishevelled tourist having cat-like mounted the ladder would, in a Uriah Heep tone, make inquiry if he might be permitted to come up. While requests of this nature were more often acceded to than refused, exceptions had at times to be made, as for instance in the case of a rowdy crew, whose unhallowed tread would disturb the just sleep of the night watchman.

As the day waxed, the crowd of visitors increased, and their thirst for information was insatiable. The names of distant peaks they had to learn, the depths of the precipices, the mode of taking observations, what was this and that instrument for, might they see through the Observatory, was the weather to be fine the following day, and oh! could such and such a fair charmer not have something light to eat? We of the staff seldom got churlish or crusty, but the perpetual interrogatory volleys from that cosmopolitan battery frequently tried our patience to the breaking point.

Few of those who reached the summit could resist the temptation, before descending, of sending a telegram, and the telegraph clerk had to be up betimes to cope with the shoal of messages which soon accumulated. Witticisms were attempted in nearly every wire, but as a rule the productions did not reach the level of even decent mediocrity. "Missed the view, and viewed the mist," while not bad, had this disadvantage that authorship was claimed for it by about 75 per cent. of those who came aloft.

Till 2 P.M., the observer on day watch saw to the hourly noting of meteorological data, but at this juncture he was called from labour to refreshment, as they say in the Masonic tongue. Hustling his comrade in science out of bed, the full complement of four all told sat down to dinner, and what a jovial meal it was! Soup *a la Beinn Neibheis* disappeared amid a surfeit of good-natured banter, then the president for the day apportioned the juicy steaks, which, when accounted for, were followed by a luscious custard and a service of fruit. Verily, good reader, we dined high on Ben Nevis, but the pleasure of our chief function was oft-times marred by the pertinaciousness of tourists who would worm themselves even into the kitchen as we regaled ourselves. "Pardong, pardong, monsieurs," a Parisian would exclaim in Anglo-French, peering round a doorway, but his further ejaculations and contortions would be peremptorily cut short by our official ejector, who would "*parlez-vous*" him in double quick time to less sacred precincts.

After dinner, the nocturnal prowler had a dog-watch or two relegated to him, which gave a respite to his companion who had been in harness since 6 A.M., but of course the latter had still four hours to account for ere his spell of work for the day was over. Mutual agreement settled this, and as the best of good fellowship prevailed, there was no quibbling over trifles. Tea came round at six o'clock, and over our cups we discussed the vagaries and plights of those we had encountered on the summit during the day. To us the visitors afforded a never-ending source of conversation, and a digest of their whimsicalities alone would furnish interesting reading.

Each evening at 9 P.M., a summary of the meteorological results for the day was telegraphed to the newspapers, and when supper had been "shifted" at ten o'clock, the staff usually participated in a game of whist. The programme varied periodically, and occasionally we had musical evenings, or perchance the fancy took us to beguile the time by reading—the Observatory having been furnished with an excellent library.

The clock never struck the hour for retiring on Ben Nevis, and although precedent had fixed it at eleven, the rule was more honoured in the breach than the observance. One cannot, however, treat repose with impunity, and so at length the man on the night-shift would be left to his own thoughts and duty. Did I say to himself? Alas! no; for during the long summer nights the stream of humanity kept toiling up the mountain steeps. Many elect to ascend in the hours of dark for the purpose of viewing a sunrise, which, if they be fortunate enough to behold, will indelibly fix itself upon their memory. In murky nights of storm, a certain unalienable eeriness naturally attached at first to the observer doing the silent vigil, but familiarity bred the inevitable contempt, and the law of cause and effect was applied in all apparent inexplicabilities.

Thus, day and night, summer and winter, for twenty-one years, observations were carried on at this station, in conjunction with one opened at sea-level in 1890, and, periodically, mid-way on the mountain at a temporary building during certain summer seasons. The Observatory on the summit was connected first by telegraph, and afterwards telephonically, both with the Post Office at Fort-William, and also with the sister station there. Two months at a time was the recognised duration of an observer's stay on the mountain-top, and on the expiry of this period he changed quarters for a like space with a colleague at the low level. During the summer months, the Directors sanctioned the

THE OBSERVATORY—SUMMER.

appointment at Ben Nevis Observatory of competent volunteer meteorologists, and these billets were much sought after, particularly by university students. The post was an honorary one to this extent that those temporary assistants gave their labour gratis; but on the other hand, they were comfortably housed, and lived on the fat of the land, not to speak of the unique and novel experience afforded by such a holiday—for so in reality it was. Many of the old Ben Nevis men are now in positions of trust and authority the world o'er, one of them, Mr. W. S. Bruce, of the *Scotia* expedition, having only recently made a bid for fame in Antarctic research, being accompanied by Mr. R. C. Mossman, who also did some valuable work on the Ben. The original superintendent, Mr. R. T. Omond, is now Secretary of the Scottish Meteorological Society, while Mr. Angus Rankin, who latterly held the office of superintendent, served on the Ben Nevis staff during the whole period of the Observatory's existence.

But for what, some may ask, has all this labour been undertaken? Of what good to any one is your mass of figures? The answer to such questions is, that when the voluminous data compiled from the observations taken on Ben Nevis come to be fully discussed, dissected, and applied, it is believed that the object aimed at by meteorologists—to wit, a recognised system of forecasting the weather—will be, if not achieved, at any rate brought appreciably nearer fruition.

CHAPTER IV.

THE COMMISSARIAT.

MANY doubtless will have wondered in what manner the Observatory was provisioned, and the curiosity is natural, in view alike of the numerous difficulties to be overcome, and the comparatively meagre means of access provided by the bridle-path. It was foreseen when the institution was established, that during the long months of winter, protracted inclement intervals would occur when it would be impossible to hold communication, except by wire, with the observers on the top, a state of things which made it imperative to lay in supplies during summer. How to do this satisfactorily and in an economic manner, was a question which for a time engaged the serious attention of the executive, but ultimately the conclusion was arrived at that the best method to adopt was horse traction.

This item alone made a big hole in the exchequer, but it was unavoidable, as the creature comforts of the staff could not be overlooked or jeopardized. The custom was to employ two horses, and when the good weather set in, these hardy animals—who soon became inured to the vicissitudes of hill life—made the ascent almost daily, Sunday excepted. Each was harnessed with a pair of panniers in which were packed all kinds of stores, supplies, etc., and it was most interesting to observe the docility and tractableness of the horses as they negotiated the angular track with their respective loads of half a ton. Provisions—chiefly of the tinned order—to serve for nine months, were taken up in summer, as well as an adequate supply of coke, which was the fuel used at the Observatory. The latter, brought up in bags, was deposited outside the store till several tons had accumulated, when one day the ukase would go forth—"All hands to the coke." This was an "extra" on the meteorological programme, but we all accepted our lot cheerfully, and that none the less because our nigger-like appearance provided gratis amusement to the visitors for an hour or two.

"Heave away!" "Slacken the guy!" "Let her go!" and such-like shouts from the throats of four begrimmed individuals, savoured more of the purlieus of a dockyard than the peaceful environment of a mountain crest on a calm summer's eve. What cared we—the task was a diverting one, and there was no fine linen or fashionable habiliments to soil. Invigorating ablutions in the dew of Ben Nevis followed; that is, if there had been no drought, for there were occasions when we were visited with a water famine on the summit. There was an excellent spring situated at an elevation a little lower than the Observatory on the south-east side, which supplied the water for household purposes, but occasionally in very dry weather a dearth of this most necessary adjunct would arise. In such instances, dependence had to be placed on a supply brought up from lower sources by the horses. Large pans or casks standing in the kitchen were filled, and so carefully was the commodity guarded that I have known the cook refuse a drink to one of the staff. Any one detected in needless waste was severely censured, and sometimes it was more politic to wear a grimy countenance than to beg a cup of water wherewith to cleanse it. In winter, of course, such a contingency never arose, for there was snow in abundance; indeed, the domestic supply consisted of melted snow, which to drink was somewhat unpalatable, although its purity was beyond reproach.

The pack horses also brought up letters, newspapers, and periodicals, and for these a rush was always made, as, without exception, we were inveterate readers, and by the nature of our occupation, probably took a keener interest in affairs political, national, and otherwise, than did those in less exalted habitations. Our convoy was sometimes subject to attack, its most bitter enemy being the cook, who, irrespective of whether he had been in office for days, months, or years, always thought it his duty to assume authority in matters of transport.

I recollect an amusing incident in this connection. The Master of the Kitchen for the time being, who happened to be an Irishman, was in the act of rating one of the horsemen for omitting to bring up a

THE LOW LEVEL OBSERVATORY, FORT-WILLIAM.

certain vegetable, when a bouncing Cockney tourist arrived on the scene and gave vent to the expression, "Here I am at last on the top of Scotland's little pimple!"

The irate cook eyed him for a second or two, and then in a tone not altogether unmixed with venom, blurted out, "Shure, sir, it's mysilf that would like to see it on your neck!"

A queer lot were the Ben Nevis cooks, albeit some of them were men of travel and culture—men who could help to make a dull day pass pleasantly. But this is a forte of old soldiers—and the majority of them were such—so we suffered them to pull the long bow so long as the action did not interfere with the due discharge of their duties. A representative of this kidney interrogated us in company once as to whether we doubted his veracity, and in reply, our chief, while disclaiming such intention, told him he had mistaken his vocation, and ought to have been a weather prognosticator! We all laughed, but the veiled insinuation was too obscure for the perspicacity of the cook.

About fifty yards or so from the Observatory stood a wooden erection where refreshments or sleeping accommodation for the night could be obtained at a tariff, which, while unquestionably high, was not unreasonable considering the cost of transit. The lessee of this place (called an hotel for courtesy), likewise employed a horse for the uptake of his stores, etc., and at special functions—some of which are referred to in another chapter—the Observatory cook waived ceremony, and set his art and experience at the disposal of our friends across the way. This hostelry remained open from June till September inclusive, during which time it was well patronized, but the young ladies in charge frequently complained, and not without reason, of being knocked up in the small hours of the morning by some belated climbers clamouring for a cup of something hot.

So long as the track was clear of snow, our own transport animals kept us well supplied with fresh viands, but there was no putting off the evil day when we had to bow to the inevitable and accept the "tinned regime" with as good a grace as possible. Winter, however, did not entirely nullify our commissariat connection. In good days when the atmosphere was clear and the snow frozen hard on the surface, the roadman, who was well acquainted with the hill, made a trip up with a budget of newspapers and letters, taking back with him missives to post and the various meteorological returns accumulated during the time we had been cut off from intercourse with the lower level. He also took up such luxuries as tobacco and cigarettes, commodities which the observers would not barter, especially during the dark days, for much fine gold; indeed, they would rather have foregone a meal a day than cease their obeisance to my Lady Nicotine.

The horse drivers were all Highlanders, accustomed to hill-climbing and exposure, and full of native wit, though unfortunately for the Southerner their anecdotes were usually couched in the euphonious language of the Gael. Speaking of Gaelic reminds me of one old character whose presence was not strange to the mountain in its ever-changing moods. This veteran, whose English was scant, was asked one day by a tourist the name of a long straggling piece of land just visible in the ocean on the western horizon. "She'll no be awfu' sure," replied the Celt, "but she'll thocht it was the Hebrews!" (meaning, of course, the Hebrides).

One of our transport retainers, known colloquially as "Dockles" caught the war fever when hostilities with the Boers broke out, and joined the Lovat Scouts—that reputable corps of Highlanders who so ably upheld the traditions of their race in the Dark Continent. Poor Dockles! he had a worthy send-off, and we all hoped to welcome him back to the Ben, but it was not to be. A Mauser bullet terminated his promising career at Quagger's Fontein, and with many another brave son of Scotia, peacefully he'll sleep 'neath the sward of Afric's veldt till the trump shall again sound the "Last Post."

CHAPTER V.

ADVENTURES.

In the ordinary sense a meteorologist's calling is one comparatively devoid of adventure, but when it is prosecuted at an elevation of a mile above the sea, the converse is nearer the truth. It was not to be expected, therefore, that the men who, in the interests of science, sacrificed so large a portion of their lives on Ben Nevis, should have come out of the ordeal scatheless. Many and varied were their adventures, some of them not unattended with considerable danger, but I shall only recount a few of the more outstanding ones.

One year, towards the beginning of February, arrangements had been made for the interchange of staff—that is, two observers from the low level station were to ascend the mountain with a view to relieve a like number of their co-workers at the summit. This phase of the scientific curriculum was, in winter, one fraught with innumerable hazards, and the greatest care had to be exercised both in going up and descending the hill. The weather on the occasion in question was apparently perfect, the sky being almost cloudless, with bright sunshine and no wind. Setting out from Fort-William at

THE PACK-HORSES—SUMMER.

10 A.M., the two relief observers drove to the base of the Ben and immediately commenced the climb.

Going up the lower shoulder, through a gully in which the path twines, the toilers were bathed in perspiration, so warm was the day, and yet, even then, a change was brewing. Over the southern brow of the mighty eminence a dense vapour-like mass was forming, and it required not the eye of a meteorologist to tell that a snowstorm was at hand. Without the slightest warning, a stiff gale, accompanied by snow, sprung up from the south-east, and the better to guard against accident, the observers at once roped themselves together. To make matters worse, the track, covered as it was with ice, became so slippery that the pick-axes had to be requisitioned and steps cut out ere a foothold could be secured. This is a tedious enough task in calm weather, but when snow and wind have at the same time to be fought against, the difficulties are multiplied.

Nothing daunted, they plodded along, although the fine powdery snow and drift, which the wind swirled everywhere, well-nigh choked them, and it was impossible in the howling tempest to hold converse, as the loudest shouts were lost to the ear. At length the flat stretch near the lake was reached and allowed of a breathing space, but no halt was called, as the half-way house towards which they were making, was close at hand. Would they make it? The elements, which by this time had culminated in a blizzard, seemed to negative the hope. Plying their ice-axes dexterously, they managed with care to make slow progress, and had even the *sang-froid* to take stock of a solitary snow-bunting weathering the storm on a boulder. At their approach, this habitant of these wilds attempted to fly off and was instantly dashed to the ground by the wind.

After many trials they reached the hut, but to their chagrin found the snow heaped in great banks round and over the structure. All attempts to force an entrance proving abortive, they cowered for a space on the lee side and discussed the situation. Only half-way up the hill, they wisely decided that it would be folly to think of proceeding, so after resting for a little they retraced their footsteps. The return journey, with the storm behind them, was less trying, but the state of the road called for great caution and presence of mind, as a false step might precipitate them headlong down a gully. The base was reached by four o'clock, and on getting home sometime later, a warm bath restored their aching and benumbed limbs to the *status quo.* Next day they set out again, and after a toilsome climb calling once more for step-cutting, and lasting over five hours, they reached the summit in safety. They were told by their colleagues, who then went down, that it was impossible with safety to venture out of the Observatory on the previous day.

The experience of an observer one dark night in September was of a different nature. He had just completed his usual round, and was on the roof taking bearings when he heard a shout, emanating, as he thought, from the hotel. Hieing thither, he ascertained on enquiry that no one there had called, and concluded that his ears must have played him a trick. Still, he was not satisfied, and on going to the door he gave vent to a long and loud "Halloo," which, to his surprise, was answered, apparently from the neighbourhood of the cliffs. Picking his way carefully in the darkness to the edge of the precipices, which here have a sheer fall of over 2,000 feet, he succeeded at length, by an exchange of calls, in coming up to two forlorn females sitting on the very brink of a cliff and almost paralyzed with fear.

Rain had been falling steadily during the evening, and the appearance presented by those two poor creatures when they reached the hotel—to which the observer safely conducted them—was pitiable in the extreme. Every stitch of clothing they wore was saturated with water, as they had been for hours in their perilous position, but after a change of garments, which, fortunately, was forthcoming, their spirits revived. Being foreigners, very little could be got out of them except a voluble effort at thanks to their rescuer. Availing themselves of the shelter provided by the hotel, they stayed all night, and took their departure, after many salaams, the following morning.

In foggy weather or darkness, the summit of Ben Nevis is a very dangerous place to those unacquainted with the topography. When the Observatory was in existence, its light could be observed from the path on the further side of the precipices, and the natural inclination was to make straight for the beacon. To do so would mean certain death, as gaps in the cliffs jut out at right angles to the road, and the latter at these points had to be carried considerably to the right, a fact which accounts for the predicament in which the ladies above referred to found themselves.

In winter, visitors were always welcomed at the Observatory, and not having been aloft for some time previous, the writer, accompanied by a friend, resolved upon a nocturnal climb in December. The night was an ideal one for such an exploit—so a wire from the top informed us, and when we commenced the ascent about five o'clock a full moon rode in an almost cloudless sky, whilst the road being frost-bound was easy to walk upon, and our spirits were high as we set stout hearts to a stae brae. All went well till we reached the tarn, but here Dame Nature began to treat us with scant courtesy.

In a twinkling, fog enveloped the hill, but this did not disturb us greatly as we meant to stick to the path, which, so far, we had experienced no difficulty in following. On we trudged, steadily mounting; but hark! what was that ominous sound? Wind, by Jove! and with power in its sough; we could hear the storm coming adown the corries. "Keep up your pluck, old chap," said I to my companion, but next moment my own courage

AT THE OBSERVATORY WELL ON SUMMIT.

fell a degree or two as a swish of snow caught our faces. Then all of a sudden we were at the mercy of the wind, buffeted, half blinded with snow-drift, and hardly able at times to get out of the spot.

When the storm broke, we had accomplished about two-thirds of the journey, and to give up then would have been galling, so we determined to proceed. Very soon icicles depended from our hair and moustaches, while the snow coalesced on the exposed sides of our faces. By this time our progress was very slow in consequence of our garments having become frozen stiff, and we could not, do what we liked, bend either our arms or our legs. Unconsciously there crept over us that almost irresistible desire to lie down and sleep, of which travellers tell us, but we fought strenuously against the temptation. The tempest shewed no signs of abating, and to make matters worse, we discovered to our horror that we had got off the path—an error easier committed than remedied.

Try as we might, no path could we find, and the awful truth that we were lost—lost on Ben Nevis, dawned upon us. Leaning on our sticks we groaned, and so utterly worn out were our bodies that it was only with the greatest difficulty we could maintain an upright position. Would that balmy sleep of death come soon, and when our stiffened corpses were discovered, who would break the news to our loved ones down below?

"Halloo!" Simultaneously our ears caught the sound, and in one voice we echoed back the gladsome shout. Soon we saw the glimmer of a lamp, and next instant one of the observers was grasping each of us by the hand. His presence seemed to put new life into us, and learning that we were only about 500 feet below the Observatory, we exerted our remaining stamina and at length reached that haven of rest. 'Twas a narrow squeak, and until the inevitable intervenes, I have no desire to obtain again so close a view of the Valley of the Shadow.

The Observatory post-runner had an experience of a kindred nature one year in February, but in his case it occurred during daylight. A favourable opportunity occurring, he was dispatched to the top with a budget of letters, etc., and the surface of the snow being hardened, he elected to take a short route up what is known as the Red Burn. Had the weather kept clear, nothing probably would have happened, but, as it turned out, a fog so thick that one's hand became indistinguishable if held out at arm's length, settled on the mountain, and of course the wayfarer at once became like a ship without a helm. His non-appearance at the Observatory created anxiety, and one of the staff set out to search for him. The latter, after some devious peregrinations, accidentally alighted on his spoor, and ultimately came up to him pretty well done up and in a position at least half-a-mile from the route which he had been following when the fog came on. Such instances shew how very deceptive the mighty Ben is, and how treacherous are the storms and mists which sweep down its boulder-strewn sides.

When going down the mountain with a visitor on another occasion, an observer was asked to initiate the former into the mysteries of the snow-slide. A word on this subject in passing. In winter, the passes and ravines soon become levelled with snow, on which, after a thaw, a glassy surface forms, and down this, sitting on a piece of wood or a stone and steering with the feet, it is possible to descend for about 2,000 feet, in a very short time. Great care and a good deal of experience are, however, requisite, in order to execute this feat without detriment to the person. The instance under review will demonstrate the truth of what I have said.

Arrived at the edge of the snow-slide, the observer was imparting information to his too eager companion, when the latter, seizing a square piece of rock, incautiously got on to the slide, and before he knew what was happening he was being whirled down the frozen declivity at an alarming pace. Small of stature, he performed, as momentum increased, some of the most grotesque somersaults and circular movements imaginable; sometimes his head would be uppermost, then his legs; anon he would cannon off a protruding wreath, rebound, cut a spread-eagle, and goodness knows what all. The while these cantrips lasted, nothing could be done by the observer to lend a helping hand, and though we laughed over it afterwards, the episode looked serious enough at the time.

A great jagged boulder stretched out across the slide some distance down, and if the victim of his own rashness had crashed against this, he would assuredly have been killed. Just as he was nearing it, the watcher, who took in the situation at a glance, lost sight of him and stood aghast. What really happened was this. About a dozen yards above the boulder, a cross burn had cut a fissure in the snow, and into this the human toboggan had providentially been deposited. When extricated, blood was oozing from his head and hands, but everything considered, his wounds were very much lighter than might be looked for in such a mad career down 800 feet of a frozen surface. I fancy he got enough of snow-slides that day to last him for a life-time.

WATER-CARRYING AFTER A THAW.

CHAPTER VI.

"RECORDS."

On account of its unique position, Ben Nevis for many years was a source of great attraction to so-called "record breakers," and numerous were the ruses resorted to, alike by individuals and companies, to secure notoriety in this direction.

Not so very long ago, a well-known firm of sewing machine makers utilized a cover page of a popular weekly periodical in order pictorially to depict one of their representatives delivering a sewing machine at the Observatory. The point aimed at, of course, was to herald the fact that the goods of the said firm were "at the very top." Sometime afterwards a London cycle manufacturing Company went one better. An agent of their's set out from the metropolis astride a motor bicyle *en route* for the summit of Ben Nevis, and strange to relate, he accomplished the feat, although many who have scaled the "misty monarch" will be inclined to doubt the veracity of the statement. Still, the fact remains, and the foolhardy cyclist secured ample proof thereof, having been photographed with his motor at the meteorological station. True it is that on his first attempt he reached only the half-way house, where, on account of punctured tyres, he had perforce to leave his machine for the night. Nothing daunted, he had the defects remedied and started next day for the summit. Endowed with indomitable pluck and perseverance, he managed to reach a height of about 3,500 feet, astride his bike, but a snowstorm and darkness intervening, he had to dismount. Not to be beaten, he shoved the cycle up the remaining stretch of rugged pathway, sinking betimes up to the waist in snow but eventually reaching the top, where he was comfortably housed for the night by the observers. Next day he descended the mountain, riding the machine part of the way, and reached the base without injury to himself and very little to his motor. A Scotch daily newspaper, commenting in an editorial on the cycle-mountaineering feat, concluded thus—" It seems doubtful if even the attempt to ride down would prove a good advertisement, even if the cyclist's obituary notice in every British paper contained an intimation to the effect that the type of cycle ridden by the deceased can now be obtained from the makers with new special adjustable Ben Nevis brake." Notwithstanding this skit, the performance was calculated to draw attention to the wares produced by the manufacturers of the motor.

When the wheelbarrow craze was at its height, the local bellman pushed one of these commodities up and down the 4,406 feet of altitude, although it must be conceded that the remuneration secured by him for his trouble was by no means commensurate to the physical energy wasted. Another has to his credit the achievement of getting a horse and cart to the hill-top, and towards the end of last year, Edge, the renowned chauffeur, was in communication with the late superintendent of the Observatory on the subject of running a motor car to the summit. The project, however, fell through—probably the best thing that could have happened—as the path was too narrow to admit of any constructed car being taken up with safety.

A foot-race from the post office at Fort-William to the top of Ben Nevis and back, was an event which for a number of years helped considerably to vary the routine of things at the Observatory. Strength of muscle and physical endurance are qualities which seldom fail to call forth admiration; but when these are employed in foolhardy and dangerous exploits, their possessor is surely acting in opposition to the laws of nature. There doubtless is a qualified degree of honour in being able to sprint up and down such a mountain as Ben Nevis in say two hours and a half, but is the game really worth the candle? What object is to be gained thereby, or what good purpose served? Verily, athletics are extending if such a departure is destined to come within the category.

No one appears to know how the mania originated, but somehow it seemed to catch on, and year after year gold medals were offered for the best record established. Some of the foremost harriers and runners of the time were lured into the contest, and the wonder is that no resultant fatalities ensued. The first authentic performance was that accomplished by William Swan, Fort-William, who, in 1895, did the 15 odd miles in 2 hours 41 minutes. Two years later, a veteran athlete in the person of Lieut. Col. Spencer Acklom, late of the Connaught Rangers, entered the field, and, while he failed to establish a new record, he evoked much interest in athletic circles by running the novel race in 2 hours 55 minutes. When it is stated that the worthy Colonel at this time was 55 years of age, it is not to be wondered at that people marvelled at his wonderful power of endurance and conspicuous development of muscle. During his sojourn in India he had ample opportunity of gratifying his love for mountaineering, and indeed scaled many of the more prominent peaks of the Himalayas.

The same year William M'Donald, Leith, made a bid for the medal and succeeded in reducing the time to 2 hours 27 minutes, but two months later Swan capped this by doing the double journey in 2 hours 20 minutes. Returning once more to the fray, the following year M'Donald again managed to take a couple of minutes off the time occupied by Swan, leaving the record at 2 hours 18 minutes. At this it stood till 1903, when Ewen Mackenzie, the Observatory roadman, greatly astonished everybody by covering the distance in 2 hours 10 minutes, a performance which is unlikely soon to be beaten.

Try to imagine what the accomplishment means.

OBSERVERS ASCENDING IN WINTER,

Fifteen miles and "a bittock" is the length of the double journey, and to travel this distance on the flat, in the record holder's time, would be no mean feat for an ordinary individual; but when one has to contend with boulders on an inclination in parts as steep as 1 in 5, matters assume an entirely different aspect. Not only is this so, but the hill sprinter had frequently to pilot himself out of banks of fog and mist, o'er mountain streams, and through blinding snow-showers, while an error in judgment might mean a broken limb or a fractured skull.

The year 1903, it is hoped, saw the closure applied to this form of sport, falsely so designated. A Glasgow harrier named Dobson, who had entered as a contestant, when negotiating the upper reaches of the mountain, suddenly collapsed and was found by some visitors in a dead faint. All the remedial measures at hand failed to revive him, and an ambulance was wired for to Fort-William, to which place he was conveyed about midnight still insensible, and little hope was entertained of his recovery. But his medical adviser, who was a "man o' pairts," did not entirely lose hope, and he had the satisfaction of bringing him round after a period of about ten hours' unconsciousness.

A local hotel-keeper also offered a gold medal for the "establishment of an authentic competitive record for the climbing of Ben Nevis," in contradistinction to the other contests in which the competitors ran singly. While one could not help admiring the pluck and prowess of the ten well-knit athletes who took part in this race, it was sad to think that they should risk their lives on the off-chance of gaining a trinket intrinsically worth not more than a couple of guineas. One or two of them spoke of the glory and honour of holding the record for the quickest journey up and down the highest mountain in the home domains of the Sovereign, but wherein lay the glory and honour if health and strength were in consequence to be permanently impaired, not to speak of the great danger to limb and life? A gamekeeper, named Hugh Kennedy, won this race, which, on account of the starting point being different, was a mile longer than in the other instances. Kennedy, who was born and bred in Lochaber, managed doubtless to outstrip his trained compeers from the south as a result of his life-long familiarity with hill climbing. His time for the performance was 2 hours 41 minutes. The gentler sex, moreover, did not escape the infection, and a race confined to them, comprising the ascent only, was won by Miss E. Tait, a post-runner, who accomplished it in half-a-minute under two hours—a very creditable achievement, considering that the average tourist takes four hours to do the same thing.

Both in literary and scientific circles, the name of the physicist who accompanied the *Challenger* Expedition—Mr. J. Y. Buchanan, Edinburgh—needs no introduction, and he it was who inaugurated a somewhat novel dynamical experiment on Ben Nevis.

Having devoted a good part of his life to mountain climbing in Switzerland and other countries, Mr. Buchanan held that the ascent of such heights, from a scientific point of view, possessed a significance which was too often overlooked. A competitor, for example, who starts at the foot of the mountain and gets to the summit in the shortest time possible, has tried his power of work doing and of endurance to the utmost, and if he only has a measure of the work done and of the time occupied in doing it, the sporting event becomes a scientific experiment of the highest value. The athletic worth of the contest is in no way affected by the measures which have to be taken in order to secure the scientific advantage. It is necessary to know the vertical height of the finishing point, the weight of the competitor, and the time which he takes to transfer himself from the lower point to the higher one.

Mr. Buchanan offered several handsome prizes, and a goodly number of competitors came forward. On the day of the race, while clear at the base, mist enveloped the uplands, a contingency which made it dangerous for those competing to leave the path, and it was almost impossible to distinguish any of them until within a few yards of the Observatory. The first prize winner was Ewen Mackenzie, already referred to, who developed exactly one third horse-power during the time he was climbing, viz., 1 hour 8 minutes.

In all these races, great pains were taken in the

matter of time-keeping, and the conditions were so stringent that no one who entered had any chance of winning a prize by shirking any part of the allotted task. While disapproving of the contest, *qua* such, we of the meteorological staff, as we could not "call it off," interested ourselves in the results, and on race days we were all agog with excitement, the man on the night watch even foregoing his quota of sleep to watch the competitors arrive and depart. Now that the Observatory has been shut up, it is very unlikely that any further such contests will be organized, and record breaking on the mountain will be as a tale that is told. *Sic transit gloria mundi!*

A ROOM IN THE HALF-WAY STATION.

CHAPTER VII.

CEREMONIALS ON THE SUMMIT.

In these latter days, when the ingenuity of man is habitually applied towards discovery, when genius would seek to wrest from Nature more and more of her carefully guarded secrets, it is refreshing to get away from the turmoil of life—from the hurry-scurry of commercialism and red tape, and seek repose for the nonce in a region where only the works of the Great Creator meet the eye. Amid these everlasting hills there is an all-absorbing solemnity—a mysterious, inexplicable fascination which seems to grip one in such a way that for the time being he loses his individuality. The transientness of all things human, the mysteries of life and death, God's omnipotence and man's insignificance—these themes, and they are only fractional, creep unconsciously into the soul, and the puny mind of the creature wrestles for a space till all becomes chaos, when, imperceptibly as they had come, the visions flit. Again, imagination takes up the thread, and ponders on things primeval; revels in pictures of ages flown, and concocts, in imagery, scenes in the early doings of our rude progenitors.

In a contemplative mood, standing alone on the summit of Ben Nevis, one cannot help wondering if its snow-clad pinnacle was wont in times past to know the tread of man, and if our forebears were in the habit of utilizing its unrivalled height for any peculiar purpose. Neither history nor legend have much to say on the point, and the assumption is that the original dwellers in these latitudes cared not for elevating themselves!

Although many had doubtless scaled its rugged sides prior to then, the earliest recorded instance of any ceremonial on the summit of Ben Nevis was in the year 1842, when Queen Victoria indulged in her first tour throughout Scotland. In a Memorial descriptive of the Royal progress, it is stated that "many were the fires which blazed on the auspicious occasion over all parts of Scotland. But the most aspiring of all was that erected and ignited by the inhabitants of Fort-William, who, with incalculable labour and perseverance, carried an immense quantity of fuel and a great many tar barrels to the summit of Ben Nevis. Salutes were fired from the ancient Castle of Inverlochy, a stronghold in which it is said that treaties were signed between Charlemagne and the early Kings of Scotland. Whilst the booming of guns awoke the echoes of Lochiel, Ardgour, Glen Nevis, and the other wild passes in the neighbourhood, the passing clouds of mist on the mountain occasionally veiled the blazing beacon; and ever and anon, as the breeze cleared them away, it burst forth with Vesuvian splendour, and shed a red glare on every mountain top around."

This coign of vantage was again chosen for a bonfire on the occasion of her late Majesty's Diamond Jubilee—22nd June 1897. It was originally contemplated to fire the pile from London by means of the telegraph wire, but through some misunderstanding with the postal authorities, this part of the programme had to be abandoned. The weather that day was a caution, snow falling heavily most of the forenoon, followed later by a dense drizzle which marred to a great extent the enjoyment looked forward to by so many.

During the day, a contingent of the Cameron Volunteers arrayed in full Highland costume and under command of Lieutenant (now Major) John Cameron, ascended the mountain and fired a *feu de joie* from the highest foothold in the land, the regimental piper thereafter contributing the National Anthem on the *phiob-mhor.* It was in a high degree incongruous at that season of the year to be standing in two feet of snow and shivering the while from the chill of stinging sleet blasts. A strong south-westerly wind swept the mountain-top and utterly frustrated an attempt on the part of a photographer to secure a plate of the loyal band who had come so high to do honour to their Queen. The company, it may well be imagined, were not loath to make an adjournment to the hotel, where they sat down to an excellent repast, after which the loyal and patriotic toasts were honoured in true Highland fashion.

The honour of igniting the bonfire was accorded to Mrs. Rankin, whose husband superintended the Observatory, and when she had gracefully performed that duty, the Bens echoed and re-echoed with rounds of cheering. Rain spoilt the effect, but had the weather been fine, the scene by night would undoubtedly have been one of the most interesting in the country, for fires were lighted on hundreds of heights which, on a clear evening, are all within view of Ben Nevis. Indeed, under favourable auspices, it would even have been possible to obtain a glimpse of any particularly prominent blaze in Ireland. As it was, only a glimmer of the bonfire could be discerned from the hotel door a few hundred yards off, and it was rather depressing after all the labour that had been expended, that that unique beacon should burn itself out to the accompaniment of a downpour of rain and an enshrouding pall of mist. Most of those who had braved the elements hurried off soon after the pile was fired, but a few remained, and we made merry for hours in the hotel, neither thinking of nor caring for the disagreeableness of things without.

Novelty was doubtless the aim of the greater number of those having to do with the Coronation festivities, and in national rejoicings of such a nature ample scope is afforded for the exercise of ingenuity. Still, when one considers the millions who took part in the gaieties, it is not to be wondered at that the contest for distinction, alike in the matter of

MIST COMING ON.

pageants, decorations, and illuminations, was fraught with keenness and rivalry, and that in order to gain notoriety, the pet scheme of an individual or community required to possess at least the special attribute of uniqueness. The bonfire ignited on Ben Nevis on the night of the 9th of August, 1902, is surely entitled to a prominent position in this category, inasmuch as it was the highest illuminated point in His Majesty's home dominions. The materials for the bonfire were conveyed on horseback to the summit during the month of June, and it was intended, of course, that the beacon should be lighted on the date originally fixed for the Coronation. Snow and adverse climatic conditions greatly interfered with the transit of combustibles, and with the elements to battle against at such an altitude, the construction of the pile at the mountain-top was a task attended with no small difficulty, not to speak of risk and danger. After a good deal of vexatious delay the cone-shaped mass was at length finished, and very much out of place it looked amid its surroundings of boulder and cairn, of precipice, and chasm, and snow-lined gully.

For weeks before the Coronation, the Ben Nevis bonfire formed the theme of many discussions, and Lochaberians felt elated that no other blaze throughout Britain would shed its aureole at a higher elevation than did the one on their own well-loved Ben. It was confidently asserted, too, that should the weather prove propitious, the spectacle from this vantage point would be a memorable one, in respect that from no other outlook would the illuminations throughout Scotland be more advantageously witnessed. True it is that some recalled with a shudder the last occasion on which a bonfire was lighted on Ben Nevis, and the vicissitudes then experienced by the spectators; but the lugubrious retrospect did not damp the ardour of a large party, who, whether the outlook proved to be good, bad, or indifferent, had resolved, on Coronation eve, to breathe less tainted atmosphere, and occupy a more exalted pinnacle than did their less favoured brethren over the British Isles.

Taking time by the forelock, the writer—who was then located below—made the ascent by night, twenty-four hours earlier than the crowd. The party of which he formed a unit numbered four all told—two visitors and two observers. We were a jolly company, and jogged along at an easy pace, discussing divers subjects, interesting and otherwise. By the time the half-way house was reached, fag was beginning to tell, and the general topic of conversation turned on the recent record established by a female post-runner who, as indicated in another place, had ascended from Fort-William to the summit in a fraction under two hours. Why! we, who were not of the gentler sex, had taken considerably longer to cover half the distance, and our ire arose. To give vent to our spleen we entered the half-way house, which was then being used as an auxiliary meteorological station, and with uncouth voices disturbed the slumbers of the two observers, who immediately sprang up in their bunks and looked dangerous, thinking they had been subjected to such treatment by some inquisitive tourist. A quick glance of recognition, however, rapidly cleared the air, and our combined laughter rang out on the lone hill-side. We had lugged up some scientific instrument for these recluses, after delivery of which, and some chaff, we recommenced our climb. The toilsome bridle-path seemed endless, and ere long we entered a thick, dank, trailing mist, which was being wafted through the gullies and o'er corry and ben by a snell nor'-easter. The cold increased in severity as we gained in height, and betimes our path lay over deep snow-wreaths which the meagre sunshine of that summer had failed to melt. At length our labours were at an end, and in the grey morning light we saw, looming through the mist, the well-known outline of the Observatory. As we entered, our sighs of relief might have awakened those abed, but fortunately they were deep sleepers, and nothing untoward happened, so we rested for a space. The observer on duty aided us to infuse a good brew of the liquor which inebriates not, and encouraged by the abundance of eatables, we did not stand on ceremony, but "fell to."

Circling round the cosy ingle, we chatted and smoked for a couple of hours or thereby, after which our friends the visitors bade us farewell, while we who remained sought refreshment in a snooze. When we awoke, the day was fully dawned, and peering out we descried with horror the prevalence of a driving mist of no uncertain density. Our simultaneous exclamation was probably more forcible than polite, still it eased our tense feelings. Was this, then, Coronation morn; this our prospect from the King's most exalted possession? 'Twas galling in the extreme, and yet we were not entirely without hope, for betimes the sun would gleam forth, and for a minute or two dispel the enshrouding veil. Our expectations, nevertheless, were rudely shaken when snow and sleet commenced to fall, and we wondered how such incongruous weather conditions would have been appreciated in London.

Interested visitors from all parts had been arriving at irregular intervals in small parties, and soon the members of the Town Council of Fort-William reached the summit. After they had dined, the Provost called on all those who had made the ascent to assemble round the Ordnance Survey Cairn, at which point he delivered a spirited address. The forward progress of the British Empire during the reign of the late Queen Victoria, of blessed memory, had, he said, been unprecedented in the history of nations, and they rejoiced that the Almighty had been pleased to spare His Majesty through his recent severe and trying illness. They were confident that under the rule of King Edward, their Empire would maintain the proud position which it held at

STARTING ON A SNOW-SLIDE.

the present day. After wishing their Majesties long life, peace, and prosperity, he moved that they forward to the King an humble and respectful message of congratulation, and he thought a more appropriate place of meeting could not be chosen for Lochaber men than on the summit of Ben Nevis—the centre of a district which at all times had proved its loyalty to King and country. The telegram sent to His Majesty, which was couched in the language of the Gael, was in these terms:—

"Chum a Mhoralachd An Righ, Luchairt Bhuckingham, Lunnainn.

"Cruinn 'an ceann a cheile air mullach na beinn a's airde 'san Rioghachd so, agus air lath' aghmhor a Chrunaidh, tha Ceannard agus Luchd-riaghlaidh Gearasdan Uilleaim le cuid eile de Luchd-aiteachaidh Lochabair, gu h-iriosal a' tairgse ceud mile failte, agus le mòr dhùrachd a' guidhe gu'n sealbhaich 'ur moralachd re' iomadh bliadhna Righ-Chathair Iompaireachd Bhreatuinn, maille ri'r Teaghlach Rioghail ann an soirbeachadh sonas agus sith.

"Cailean Young,
"Ceannard Luch-Riaghlaidh a Ghearasdain."

The translation whereof is as follows:—

"To His Majesty the King, Buckingham Palace, London.

"Assembled together on the summit of the highest mountain in this kingdom, and on the auspicious day of the Coronation, the Chief Magistrate and rulers of Fort-William, with some other inhabitants (or citizens) of Lochaber, humbly offer a hundred thousand congratulations, and, with great good will, pray that your Majesty may occupy for many years the throne of the British Empire with your Royal family in prosperity, happiness, and peace.

"Colin Young,
"Chief Magistrate of Fort-William."

After the telegram had been read, the healths of their Majesties were pledged with Highland honours in and amid the "dew of Ben Nevis."

On the same day a highly interesting Masonic function was performed within the Observatory by Lodge Fort-William No. 43, the transactions of which shew that the brethren have always been conspicuous for their loyalty. To live up to this tradition, the members determined not to be less zealous than their forefathers in celebrating such an important event as the Coronation of him at whose birth their progenitors had so enthusiastically rejoiced. Accordingly, after taking a prominent part in the fêtes at Fort-William, they made the ascent of Ben Nevis, and at the domicile which stands unrivalled in Britain in the matter of altitude, opened their Lodge, when a goodly number of candidates, including one of the observers, were initiated into the mystic Order of Freemasonry.

Those members can with impunity lay claim to the distinction of having been initiated on Coronation Day, on the highest point in the realms of that Royal personage, to honour whom they had that day assembled. After repast and toast at the hotel, the Masonic ceremony was fittingly concluded by the despatch of an appropriately worded telegram to the King.

Dusk by this time had begun to set in, and interested parties were asking each other when the much-talked-of bonfire was to be lighted. The mist had never entirely lifted, and to make matters more disagreeable, rain was then falling in heavy showers, making the prospect—such as it was—one of the bleakest and most dreary that could well be imagined. Nevertheless, the beacon, which, it was anticipated, would serve in the manner of the old-time fiery cross to others assembled for a like purpose on lesser heights, would require to be fired, and at 9 P.M. this duty was carried out by Major John Cameron, as representative present of the Superior. At the Observatory, only the glare could be seen, and it seemed somewhat of a pity to consume amid that region of snow and mist, that unique representation of so much labour and money. When the combustibles were fairly ablaze, a series of fireworks and rockets were let off, but this display was alike futile, and not a few heaved a sigh of relief that the day's functions were over. An adjournment was made for a short space to the hotel, where some more "dew" was encountered, after which the company left the summit, followed with a cheer from the observers, and a few others who chose to remain. Whist, song, and sentiment prevailed at the hotel long after the guests had departed, and the writer will not soon forget the vicissitudes, pleasures, and enjoyment experienced by him on Coronation Day at the highest house in the land.

THE START OF ONE OF THE MOUNTAIN RACES.

CHAPTER VIII.

ACCIDENTS.

Comparatively speaking, the dwellers on Ben Nevis enjoyed a remarkable immunity from accident; but looking to the nature of their vocation and remoteness of habitation, it was hardly to be expected that, being in perils oft, they should emerge from the fray without a scar. Hair-breadth escapes they had many—superstitious fears none, although one member of the staff avowed that his life was preserved through the sagacity of a collie dog. The incident, which has in it an element of the much maligned "second-sight," is worth relating.

The observer referred to had just concluded his furlough at sea-level, and was on his way to the summit to relieve a mate at the upper station. When about a mile on his road, he was joined by a collie belonging to a relative, and, naturally enough, at once ordered the animal home. To all appearance the dog complied with injunctions, but when the observer reached the point where the bridle-path commences to ascend the mountain, he found his canine friend waiting for him. Chastising it with his stick, he again peremptorily commanded it to return, and as the collie scampered off in the right direction, he concluded that at last it had gone home.

When tackling the long, steep ascent in the corry of the lower shoulder, a further surprise awaited the observer, for, in rounding a quick turn, he once more noticed the dog in front of him. The pertinacity of the animal troubled him a little, and after communing with himself, he decided at last to allow it to accompany him—a wise decision, as the sequel will shew.

All went well till the pair came to the half-way house, when climbing became dangerous, owing to the upper part being entirely covered with snow having an icy surface. Walking very cautiously, the observer struck off the road and followed a direct course to the Observatory *via* the Red Burn. The better part of the ascent had been accomplished, and above him the relief man descried two of his colleagues coming to meet him, but they, on seeing him so close, returned to the Observatory. Putting on a spurt with a view to overtake them, he accidentally lost his footing, and, of course, was soon being precipitated down the glazed snow at an ever-increasing momentum.

It was at this juncture that the sagacity of the dog was displayed, for, as soon as he saw his companion sliding, he seized his coat with his teeth and held on tenaciously, thereby helping in no small degree to check the downward impetus. But on they went, sometimes tossed over a protruding rock into the air, next hurled head foremost through passages of ice and snow, until at last, when tearing at a terrific rate down the gully, a rise in the ground offered the only chance of escape. The observer, not slow to seize the opportunity, with some dexterous movements in which the dog intelligently joined, steered for the bank of ground, up which he and his faithful companion—the latter still holding on to the coat—were brought to a stand-still, after a miraculous escape, having covered in their downward slide a distance of over 2,000 feet.

He was considerably bruised and pained in every limb, and the dog's paws were all shewing red flesh; but the injuries sustained, looking to the hazards, were relatively slight. Though greatly exhausted, the observer forced an entrance into the mid-hut, and here he and his dog fraternized until discovered by a search party who set out from Fort-William on learning that the scientist had not reached his destination.*

* Since these sheets have gone to press, an accident, similar in many respects to this one, but attended with more serious consequences, befel the Rev. Mr. Robertson, Edinburgh. A prominent member of the Scottish Mountaineering Club, he had been indulging in a climb on Ben Nevis early in April, and when descending from the summit on a snow-slide during a thunder-storm and blizzard, was rendered unconscious, by what agency he could not say, and hurled ruthlessly down about a couple of thousand feet. He could remember nothing till he found himself making for the track on the lower shoulder of the hill, and opined that he had been rendered insensible by a

During a severe thunderstorm in June, 1895, the Observatory narrowly escaped destruction by fire. For hours the lowering sky presaged some coming disturbance in Nature; snow and hail fell alternately, and a semi-darkness settled down on the hill-top. The needle of the telegraph instrument kept continually clicking as the result of earth or electrical currents, and the stove-pipe anon emitted sparks and smoke. About 3 P.M., or as the log-book has it, at 14 hours 57 minutes, a blinding flash of lightning illuminated the Observatory, followed instantly by a terrific crash. Bluish spurts of flame and a cloud of smoke burst from the telegraph instrument, and the cook who was sitting in the office at the time, was pitched on his back and rendered unconscious. Fortunately, beyond a shock to his system, he was not much the worse, and his state of insensibility was not of long duration. It was otherwise with the telegraphic apparatus, which was completely wrecked, some parts of it being fused and melted beyond recognition. One of the observers afterwards purchased it from the Government, and it is now in the

flash of lightning striking either his ice-axe or a rock near where he fell. He was lacerated and seared with gaping wounds on the face and head, while all his limbs were bruised and full of sores. Minus cap and ice-axe, and with rent garments, he managed, in a semi-comatose state, to reach his hotel in Fort-William, but how he did it is almost as great a mystery as his marvellous escape from death.—*Ed.*

SNOW DISAPPEARING FROM THE SUMMITS.

custody of the Scottish Meteorological Society as a memento of elemental fury on Ben Nevis.

In the kitchen, the same flash hurled a large flower tin, a box, and a picture across the room, while the wires, which enter through the apartment, were fused in different places. Soon after, smoke and flame were seen issuing from behind the wainscotting separating the kitchen and the office, and the alarm being raised forthwith, all hands were called to extinguish the fire. The interior woodwork of the building being as dry as tinder, the odds of successfully coping with the outbreak were against the observers, who, although they had pitted themselves against the elements in other garbs, had never previously fought against fire.

While two of the staff kept saturating the affected parts with water and snow—which latter then lay deep—another was engaged in transferring the meteorological records and data outside, in case of the flames gaining the mastery. The worst fears, however, were disappointed, and after a spell of probably the hardest manual labour undertaken at that station, victory in time lay with the observers. The exciting incident formed the subject of conversation for days afterwards, and disorder reigned in the erstwhile peaceful abode where regularity and routine had not before been so rudely disturbed. The dislocation of telegraphic intercourse stopped for the nonce the transmitting of the usual reports to the newspapers—a phase of the work which had to be abandoned until a new telegraph instrument was fitted up a day or two later.

To the unwary, pitfalls abounded on Ben Nevis, and death lurked even in the pleasurable exercise of a favourite pastime. When deporting himself on ski one fine afternoon, a temporary observer—Wilton to name—met with a somewhat serious accident. Not content with prosecuting the exercise in parts that were comparatively safe, he must needs give the pastime a more exciting turn, and elected to indulge in a jump, after the manner of the expert Swiss. Following up his resolution, he glided with *sang-froid* to a shelving part of the steep, but just as he neared the brink, one of the ski becoming loose, he lost his equilibrium and buried himself head foremost in the snow at the foot of the leap.

It was well that another member of the staff accompanied him, or a very serious tale might have had to be told, albeit the episode was grave enough at its happening. When extricated, the man of ski was unconscious, and it was no easy task for his mate to bring him round. A hardy fellow all the same was this same Wilton, and ere long he was strapping on his foot-gear once more for another bout. Obstacles to him were only created to be overcome, and when fully into the swing of his sport, he managed successfully to negotiate the jump which only a short time previously had well-nigh terminated his existence. A great traveller, he was one of those who accompanied the Scottish National

Antarctic Expedition—which, as mentioned elsewhere, was under the leadership of an old Ben Nevis man—and it was in the Arctic regions when attached to the Jackson-Harmsworth Expedition that he acquired his knowledge and love of the interesting pastime of ski-ing.

In the discharge of their duties, and especially during boisterous weather, the observers had numerous near shaves of being blown over the cliffs. In fact, such a contingency was more than once averted simply because the precaution of roping each other together had been resorted to before emerging from the safety of the Observatory. Not infrequently, when reading the outside instruments, with a gale on, these weather-watchers were knocked over like nine-pins, and had to hold on to the boulders for dear life, crawling on hands and knees to the house in lulls of the storm. Often were they injured and sorely bruised in this wise, while at times, flying pieces of ice or frozen snow forced from the hill-top by the wind, contributed also to their list of wounds. Not to be despised was this latter danger, for the missives were hurled broadcast with extreme velocity, to such an extent, indeed, that the screen of a thermometer-box was once demolished, and a window broken, from this cause.

The saddest part of this short history now comes to be written. Between ten and eleven o'clock one forenoon in September, a visitor called at the Observatory and reported that there was some one lying at the side of the path near the first gorge, either dead or unconscious. The man who had been on night watch was awakened to take the next hourly readings, while the other observer and the cook went to investigate the matter. In a brief period the latter returned, and with tears streaming down his cheeks, apprised the then only occupant of the Observatory that the victim was Duncan Macgillivray, the telephone clerk. A stretcher was hastily improvised, and the twain made for the spot where the melancholy discovery had been made. In a recumbent position, a little off the pathway, lay all that was mortal of poor Duncan, his countenance lit up by an expression of calm and peace, as if grim Death had merely brought repose after toil on the rugged steeps.

Slowly and sadly he was borne to the house where but two nights before his merry laugh and lightsome banter had made him the life of the little company. On his own bed we laid him, and with the gentle drawing to of the door, an overwhelming sense of loneliness and gloom seemed to pervade the whole precincts. In subdued whispers we spoke of the life that had been; and how unspeakably sad was it, every hour, to pass by the death-couch of one who had been such a favourite in that selfsame abode.

Having descended the Ben on a Friday to take part the following day in the annual shooting competition of the Cameron Volunteers at Fort-William, of

A CORNER OF ONE OF THE PRECIPICES, SHEWING OBSERVATORY.

which he was a member, Duncan had arranged, if Sunday were good, to return on that day to the Observatory. Faithful to his promise, he commenced the ascent in the evening of that day, and at intervals, as the shades began to fall, those on the top went out in turn in anticipation of meeting him, but in every case they had to return disappointed. No intimation that he had left was wired to the summit, otherwise steps would have been taken to guard against any mishap to him. As matter of fact, he had reached the top—probably sometime in the hours of dark—and the spot where he was found was within hail of the Observatory. The grief of the observer who had been on night watch was in a measure more acute than that of the others, as he felt that, had he taken a short walk down the path after each observation, the fatality might have been averted.

A medical man who accompanied the police authorities and others to the top on the day the body was found, certified that death had ensued from syncope. In the dim twilight of that autumn afternoon, the remains were laid on a stretcher and conveyed to Fort-William for interment. The transit of a corpse by such means is a painful task at all times, but when the gruesome duty has to be undertaken in darkness adown a bleak mountain-side, the sadness is doubly intensified, and none who took part in that melancholy procession of death will ever forget the sorrow and gloom to which it gave birth. To more than one heart recurred the lines in a doleful pæan of the district—

"Spring will return over mountain and glen,
And the wilderness blossom in beauty again ;
The linnet will carol his song as of yore,
But we may return to Lochaber no more."

CHAPTER IX.

PHENOMENA.

THE phenomena, meteorological, optical, and electrical, to be witnessed on Ben Nevis are varied and of the most interesting nature. The result of special study given to this subject by the observers, forms no inconspicuous part of the Transactions of the Royal Society of Edinburgh, and many of the observations which follow have been culled from this source.

As the observers were practically in the clouds for most of the time, many opportunities were afforded of minutely examining the optical effects of mist or cloud on the rays of the sun or moon. The most interesting sights of this class were seen during moderately fine weather in winter, when the hill-top was clear of all dense fog, but the atmosphere not too dry. When a thin, almost imperceptible, film of scud-cloud or mist covered the moon, coronæ of the most vivid colours were formed. These coronæ, as is well known, consist of coloured rings arranged concentrically round the moon or sun. Each ring has all the usual spectroscopic or rainbow colours to more or less perfection arranged with the red belt outside. A very curious and not uncommon type is one in which there is a well-marked red ring with yellow and blue inside, but with also a blue margin or glare outside it. The colours inside, and including the red, make up the usual spectrum, and this margin is a kind of extra and unbalanced development of blue.

Another allied optical phenomenon is that known as "Glories." In winter, when the sun was low, even at noon, the shadow of a person standing near the cliff that runs all along the northern side of Ben Nevis, was cast clear of the hill into the valley below. In bright winter weather this deep gloomy gorge was often full of loose shifting fog, and when the shadow fell upon it, the observer saw his head surrounded by a series of coloured rings, from two to five in number, varying in size from a mere blotch of light up to a well-defined arch six or eight degrees in radius. This phenomenon did not present quite the same appearance as the better-known "Brocken Spectre," for here the shadow of the observer, in consequence of the distance of the mist from him, did not seem unnaturally large; in fact, the image of the head appeared as a mere dark speck in the centre of the coloured rings. These "Glories" were less common in summer, though they have been seen near sunrise and sunset. The sun is too high at noon to cast the shadow far enough, and the atmospheric conditions are not so favourable to the formation of the necessary fog in the valleys. The exaggerated size was occasionally noticed when, towards the setting of the sun, the shadow was thrown on very thin fog.

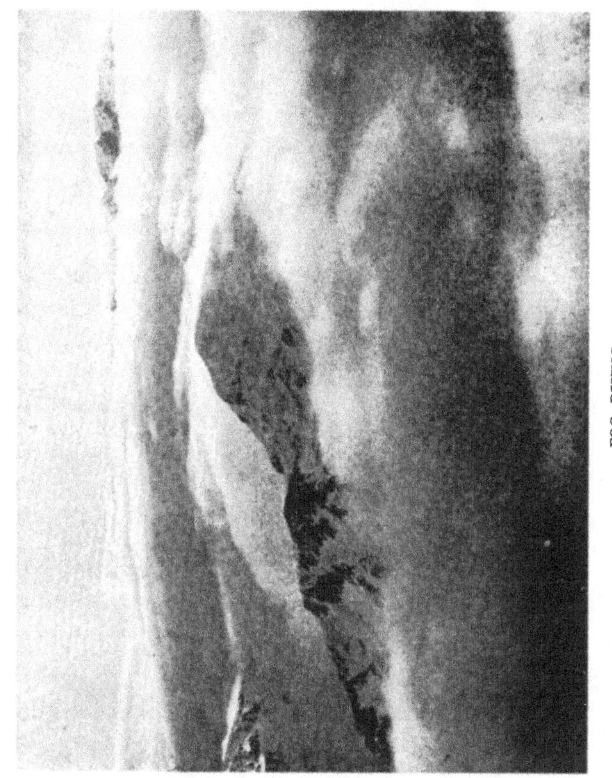

FOG RISING.

It then formed a long dark tunnel running back into the fog, and appeared further away than it really was—the necessary condition for apparent enlargement.

Rainbows were seldom seen from the Ben. During the showery weather required for their formation, the summit of the hill was usually shrouded in mist, and the view limited to some twenty yards. Fog-bows, however, were frequently observed, both solar and lunar, being formed on the loose fog driven across and around the hill when the top was clearing after a misty day. The laws governing the arrangement of colours in these bows and in the "Glories" have not yet been thoroughly grasped by scientists; in "Glories" the red side of the spectrum always appears outside, in fog-bows it may form either the outer or inner edge of the arch.

Several Halos of interest have been seen from the Observatory. The more common were of 22° and 45° radius, but on one occasion a part of the rare arc of about 90° was observed, and this is believed to be the fourth instance on record. Contact arches touching the ordinary halos, and mock suns, either white or coloured, have likewise been witnessed, as well as the vertical pillar passing up or down from the sun's disc. Once this latter form was noticed passing down between the Observatory and the hills at the other side of Glen Nevis, shewing that the snow-crystals forming it were less than three miles distant, and floating below the level of the Ben.

"St. Elmo's Fire," an electrical phenomenon of peculiar interest, seen to advantage and with frequency on the summit, was much commented upon and discussed by the staff. It usually appeared like little jets of flame on the lightning-rod, anemometers, etc.; but in the more brilliant displays, every post and chimney was tipped with fire, while sparks glimmered on the observer's hat, pencil, or fingers. An uncanny hiss or buzzing noise always accompanied it, and almost invariably it was the precursor of a snow or hail fall. One July evening, the observer, on going out at 9 P.M., had his attention drawn to a high post which was sounding like a telegraph pole carrying a "noisy wire," and on turning his face upwards he felt a gentle pricking sensation all over it. It was too bright to see any of the sparks which, under favourable conditions, would doubtless have shewn numerous and varied, but the peculiar sound lasted for about a quarter of an hour.

When a thunder-cloud approached Ben Nevis, lightning was usually to be seen flashing from it, but when it enveloped the hill-top no lightning was discerned, the alternative being a downpour of rain or a heavy fall of snow. As the cloud moved off, a discharge took place, not merely from the cloud itself, but from all large metallic bodies in the Observatory. Brilliant flashes sprang out from the stoves, and a sharp crack like a pistol shot was nearly always heard, while almost every member of the staff, at one time or another, received shocks.

Fog-crystals presented a most interesting and fantastic spectacle on the summit, and visitors were wont to marvel alike at their beauty and peculiarity of formation. In winter, when the temperature was below freezing-point, the effect of fog was to cover everything with long feathery masses of crystalline snow. It seemed that the wind-driven fog no sooner brushed against any obstruction than the moisture in it condensed in minute specks of snow or hoar-frost. Gradually these accumulated until long cone-shaped crystals, pointing to *windward*, were formed, and, by continual accretion, these grew and grew till their weight was so great that they broke off. These crystals have been seen to grow to a length of about six feet, and a diameter of half that measurement, the nucleus of the whole being a simple wooden post, some six inches by three in section. This, of course, was the result of several days' growth, and with the shifting of the wind, the spear-like protuberances formed at different angles, making the effect in a glinting sunlight one of the most artistic of the many beautiful sights to be witnessed from that coign of prospect.

The aurora borealis was oft-times seen to advantage from Ben Nevis, and on a clear frosty night, with star-spangled sky, the flitting and recurrence of these Northern Lights, as they shot up from horizon to zenith, never failed to claim the attention of the observers, who, for long intervals, viewed them in company from the outlooks of the tower. The

Zodiacal Light, a kindred phenomenon, likewise came in for much discussion at the Observatory; but the subject being still a controversial one, and there being nothing to go upon but hypothesis, brain effort in this direction was, by common consent, deviated. Sir J. Norman Lockyer defines the so-called Light as "a ring of apparently nebulous matter, the exact nature and position of which in the solar system have not yet been determined." It is usually seen as a well-marked sloping pyramid of greyish light, and was frequently observed on Ben Nevis in the east before dawn in spring, and in the west after dusk in autumn. Unless familiar with the phenomenon, a person is very apt to look upon it as merely a prolongation of twilight in the evening, and a forerunner of dawn in the morning. Indeed, there is a proverb among the Indians to the effect that "if the dawn is seen and goes to bed before he gets up, it is sure to rain before soon." Amongst thinkers, this apparent paradox is believed to have reference to the Zodiacal Light.

Only one instance—and that a very doubtful one—of an earthquake having been felt on Ben Nevis, has been recorded. Towards the end of 1884, while one of the observers was using a telescope in the tower, he felt several distinct shocks or tremors, more intense than, and of a different character from, the vibrations caused by wind, and the occurrence was entered in the log-book as an earthquake. Seismic disturbances in the surrounding glens are not

DESOLATION.
Loch Eil seen in the distance is 10 miles in length.

uncommon, and there is nothing unnatural, or far-fetched in the idea that a perceptible tremor might be experienced on the hill-top.

Another peculiarity, not unattended with weirdness, which has doubtless impressed climbers who found themselves early on the mountain peak, is the shadow of the mighty Ben. Towards 4 A.M. on a summer's morn, as the sun emerges from the rolling fog banks, the shadow of the lofty mountain is silhouetted clearly on the western horizon at an apparent distance of 30 or 40 miles, but as the orb of day mounts in the heavens, the far-off patch of black slowly sinks, and finally loses existence in the gleaming waters of the Atlantic.

A remarkable climatic feature in regard to temperature is also worthy of note. In numerous cases during winter, thermometrical readings on Ben Nevis were ascertained to be considerably higher than those taken at sea-level—that is to say, it might be freezing at Fort-William, while a decided thaw prevailed on the summit. The most probable explanation of this is, that the air, which is practically void of water, vapour, and dust, in its downward course, has no opportunity of gathering any till it reaches the surface of the ground, for all the vapour and dust in the air ascend from below. Hence, the air comes down to the mountain summits comparatively warm and very dry, but as soon as it brushes the surface of the ground, or comes into contact with air that has been lying near the ground, it begins to absorb moisture and gather dust. It follows, therefore, that the heat required to evaporate the water, which thus saturates the lowest stratum of the air, will be largely taken from the air itself, which is thus left cold and damp.

Another partial reversion of natural laws was detectable in the phenomenon designated "Silver Thaw." This peculiarity, which consists of rain falling when the temperature is below freezing point, and congealing as it falls, was of common occurrence on Ben Nevis. A prolonged fall of silver thaw occasioned considerable inconvenience to the observers; the rain froze on their coats, gloves, and even on their faces, while everything outside became covered several inches thick with solid ice, but the most serious effect was the choking of the louvres of the thermometer screens, necessitating continual changing of the boxes.

The freaks of wind on Ben Nevis often baffled lucid explanation. For example, at that altitude, it might be blowing a perfect hurricane—say from 70 to 90 miles an hour—whereas, at Fort-William, the waters of Loch Linnhe would be unruffled by the tiniest zephyr. The reverse, to a lesser degree, sometimes held good, and though the observers at the summit station might be basking, amid a dead calm, in the warmth of a glorious sun, their colleagues below would be shivering from the effects of a piercing east wind, untempered by any beams from old Sol, whose countenance on such occasions was usually obscured by dense clouds.

CHAPTER X.

STORMS.

To thoroughly appreciate a Ben Nevis storm, one would require to experience it, for any attempt at pen portraiture can give but a feeble notion of the majesty and might of the wind as it sweeps unchecked with tempestuous roar over the summit of this peerless height. With no intervening object to break its force, the true velocity of a gale was always felt at the Observatory, but so enormous was the pressure at times, that anemometers were absolutely useless as a means of registering the speed.

During the existence of the Observatory, an approximate scale was employed in estimating the force of the wind, this being represented by the figures 0 to 12 - 0 = calm, and 12 = hurricane. Force 10 to 11—which might mean anything from 90 to 140 miles per hour—was the greatest ever experienced on the top, and, of course, when the equivalent of these figures was reached, no one dared venture outside. Force 12 was never touched, this value having been reserved for use when the roof was blown off the Observatory, or the observer hurled over the cliffs! In the latter case his survivor would make an entry in the Log-book to this intent :—" Observer blown over cliff; wind force probably 12." Neither of these contingencies, fortunately, arose, so that there is no entry of 12 in the meteorological records, nor a note like the above in the Log.

In the study of storms, and weather in general, the first place is usually given to the data furnished by the barometer. To an observer at a fixed station on the *qui vive* for hurricanes, the absolute reading of the barometer at any hour is of small importance. Whether this reading be high or low, it does not help much, for storms may arise when the barometer is high, and fine weather prevail when it is low. The point carefully to be noted is the rate at which the mercury falls or rises. To illustrate this, I may be permitted to recite the conditions precedent to a memorable storm which occurred on the 22nd of November, 1898.

Till 3 o'clock on the morning of that day, the barometer was actually rising; thereafter it began to fall, and while the wind was N. or calm, the rate of fall was slight and fairly steady. In the afternoon, however, when the wind commenced backing to S. E., a decided fall set in, the rate of descent rapidly increasing, till at 10 and 11 P.M. it was as much as a tenth of an inch per hour. Curiously enough, this rapid rate ceased at midnight, and during the rest of the storm the barometer fell very little, nor did it shew much of a rise when the wind moderated.

Up till 9 A.M., the wind was a very light one from the north, followed latterly by a dead calm; but at

ON THE OBSERVATORY ROOF—A FINE CLOUD EFFECT,

1 P.M., a breeze sprung up from S.S.W., from which point it veered and backed spasmodically. Gradually increasing in volume, it reached the velocity of a gale at 9 P.M. and continued with unbated fury for several hours, blowing probably at the rate of 2½ miles a minute. Just before the wind increased to a gale, snow began to fall, and continued descending heavily till 7 A.M. on the following day. Showers had been falling for some days previously, so that there was a moderate coating on the hill-top at the time the gale came on. The manner in which this old snow, and that falling at the time, drifted during the height of the storm, is difficult to describe. The temperature having been so low then, and for a few days before, the snow was almost as dry and fine as dust. In the strong wind, this snow-dust mixed with the air like smoke, and was forced into the Observatory through the tiniest chink. This propensity of finding its way through everything and anything, taken along with its suffocating properties, made this species of drift one of the most disagreeable factors with which the scientists had to contend.

The lower doors were tightly closed and bolted early in the gale, but in the morning there was fully a ton of snow in the lobby and passages, while the bedroom and kitchen floors were covered to within a yard of the stove. Inside the Observatory, during a storm, life varied in comfort according to the time of year. In spring, when the maximum depth of snow was recorded, the rooms were extremely cosy;

but in an early winter storm, with a low temperature, the conditions were otherwise. When the storm in question broke out, the walls of the house were in their state of summer nakedness, with the result that wind and snow were forced through, to the decided discomfort of the indwellers. On the morning of the 23rd, although the stoves were kept blazing, the temperature in the office fell to 27°; and in the kitchen, a thermometer hanging on the wall six feet from the fire, dropped to 2° below freezing point.

Of course, it was impossible to face such a storm, and several of the hourly observations had accordingly to be abandoned. The internal state of matters likewise put sleep out of the question, and there was nothing for it but to make the best of a bad job, which was done by spinning yarns around the ingle, the time our bogie-roll vanished in smoke.

Now, such a storm as I have tried to tell of, would have played havoc in the lower regions. Timber would have been felled by the acre, house property destroyed, and the telegraph system disorganized, capped with a gruesome tale of innumerable shipwrecks, but on Ben Nevis it was different. Here, in the daytime, at the season spoken of, there was always the same blank whiteness above, below, and around, with not a single particle the eye could rest on, and at night an awful pitchy darkness which might almost be felt. There was the terrible boom of the gale on the sides of the hill, the rattle and groaning of the chimney pipes, the constant vibration of the tower, and the ever-present feeling that the rest of the universe was tearing past at double the speed of an express train.

A glance, *en passant*, at a snowstorm in these wilds. While an additional height of a few hundred feet would have placed Ben Nevis in the perpetual snow-line, the monarch of mountains is never entirely denuded of the white element. Time was—at least legend has it so—that the proprietory rights of the hill were regulated by snow. It is alleged that, by an ancient charter, the mountain was to revert to a prior owner if at any time the snow utterly disappeared from the gullies, and on an exceptionally hot summer the only white patch remaining had deteriorated to such an extent that retainers were dispatched with materials to protect it from the incessant rays of the sun. Whether this tale be fiction or no, it is probably safe to say that there is no one in the present generation who can truthfully say that he failed at any time to discover some snow on Ben Nevis.

An awesome thing is a storm of snow when encountered where Nature has free play! From an outlook such as Ben Nevis, the silent white volume could be seen approaching when miles off, and an entrancing sight it was to watch the varying effects athwart the caliginous glens and fastnesses. Fanned at first by gentle airs, its progress, though slow, was unmistakeable and sure, until at length the hill-top was included in the affected area. Blotted out then was the pleasing scene, and imperceptibly the nearer objects also became obliterated as fog came down to play its part in the drama. As the fall increased, so, invariably, did the wind, and 'twas then that things began to hum in these isolated uplands. Owing to the keen frost which usually prevailed, the snow, as the wind caught it up, was converted into the minutest particles and driven broadcast in dense, blinding clouds. As already indicated, conditions such as these, were abhorred by the observers, and for this they had substantial reasons.

One dark night, when things were as described, a temporary observer who had not previously experienced this phase of weather, was taught a lesson which he seriously took to heart. He was accompanied to the tower door by the permanent official, and warned before going out to grasp the guide-rope. Disdaining to act on this sage advice, or perhaps to shew his grit, he leapt on to the flat roof, and was at once hustled whence he would not by the wind. Losing his bearings, he failed utterly to make the door, and becoming soon almost asphyxiated with drift, he feebly shouted for help. His mate, better acquainted with the "lie of the land," made several unsuccessful attempts to locate him, but eventually seized him just on the edge of the Observatory wall, when he quickly hauled him to the safety of the tower. The rescued man could

CONSPICUOUS FOG CRYSTALS.

not speak for a space, and no wonder, for his entire features were covered and sealed with a mass of frozen snow particles, while his apparel, even to the semmit, was thickly intermixed with drift.

How brief a period was requisite for the formation of wreaths! A single night was sufficient to alter the whole contour of the summit, and in the same time the Observatory would be buried in the enormous accumulation of snow, which formed into such huge drifts that the roof of the building became practically level with other parts of the hill-top. In this lay the security of the domicile, for as soon as the snow hardened round about it, the warmth was retained, and outside influences affected not the interior. On the brinks of the precipices, the snow was drifted in cornices for many yards outwards, and the background being white, this, in winter, proved a veritable death-trap to the unwary. Some of these cornices had a depth of from 15 to 25 feet, and in sunshine they were most beautiful to gaze upon. What an aspect of change was presented after a night of stormy snowfall!—gorges and crevices were filled up, ravines obscured, and wreaths—nothing but wreaths—existed everywhere. Only the guide-posts marked the course of the pathway, and perchance a vestige of the hotel might be descried protruding through the dazzling whiteness. The greatest average depth of snow registered at the summit was from 12 to 14 feet.

and in the matter of temperature the lowest reading was within half a degree of zero.

"It never rains, but it pours," though a quaint saying, is applicable enough to Ben Nevis. Some wiseacres in referring to rain speak of it as a deposit of moisture. Ah me! would that they stood in a shower on this elevation of the Grampians. The water there literally comes down in sheets, and at times the saturation would be quite sufficient to wash away a fair-sized village. Why, during one day, in October, 1890, nearly 7½ inches of rain fell on Ben Nevis, and a single month in the same year was credited with 48 inches. The greatest annual fall recorded was in 1898, when the deposit reached the remarkable total of 240 inches. To the ordinary individual, these figures will convey but little meaning, but when it is borne in mind that one inch of rain represents 100 tons of water to each acre, the abnormity of the so-called moisture will be more readily grasped. Assuming the mountain summit to be about ten acres in extent, this would indicate that, in one year, the water poured down there reached the colossal total weight of 240,000 tons.

When one ponders upon the possibilities of such an inherent water power, it is not to be wondered at that many schemes are projected for the utilization of this dormant factor. Why should Lochaber, precluded on account of configuration and climatological conditions from participating in arable pursuits, not take advantage of the untold wealth represented by the waters at present unharnessed? Were any such scheme to reach fruition, it would, to a large extent, settle the vexed question of rural depopulation, and help to bring about the advent of that new era for the Highlands of which philosophers have so often dreamed.

CHAPTER XI.

PROSPECT.

It would require a person with a fecund imagination to concoct mentally a scene more entrancing or impressive than the prospect from Ben Nevis in the glorious days of summer. Lakes and streams innumerable, rolling moorland wastes, rivers, copse, corry, and tarn, flit before the vision, and vie with each other for predominance. Now and then, mayhap, a thick bank of cloud or fog gathers in the surrounding valleys, leaving only the hill-tops visible, and immediately a transformation in the scene occurs. To all appearance, one is gazing on a wide expanse of ocean studded with countless islets, some of which, as the film rises, gradually disappear, as if buffetted and over-washed by surf. In closer proximity are yawning chasms and steep declivities, the eternal snow ridges contrasting strongly with the dark basaltic rock forming the angles of the precipices, while the prolixity of colours, the murmur of the breeze, and the sublimity and peace of all around, speak to the heart as nothing in the lower world can.

Sunset and sunrise have their own peculiar significance and beauty, but of these, tongue cannot adequately speak, nor painter depict. In the sub-

FOG CRYSTALS.

dued hush of early morn, the sombre grey of the eastern sky, as day waxes, gives place to the effulgence and glory of the rising sun, which, in turn, tinges with gold the wide expanse of towering heights, dispelling the mists from the valleys, and casting a sheen of splendour on the waters of the distant ocean, where sea and firmament seem to meet on the far western horizon. Again, in the long twilight and afterglow, when the orb of day has run his course, and dips gracefully from view, the spectacle indeed is sublime; and as the eye takes in the picture formed by the saffron and roseate-tinted clouds, the jutting promontories, and the serrated mountain ranges, silence is the only tribute of admiration that one can bring.

Some reference to the more prominent hills, etc., observable from the summit, may not be out of place, and readers by consulting the appended map in conjunction therewith, will the more readily familiarise themselves with the general topography.

To the northward, looking across the deep gorges and side spurs of Ben Nevis, a comparatively level plain is seen extending from Fort-William to the entrance of Glen Spean, up which glen, to the N.E., a glimpse may be had of Loch Laggan, the background being filled in with the massive ridge of the Cairngorms, among which, Ben Macdhui, the second highest hill in Britain, can, with difficulty, be distinguished. The northern side of Glen Spean is formed by the rounded masses of the Monagh Lea hills, running up into which Glen Roy, with its famous "Parallel Roads," may be noticed; and due north from Ben Nevis lies the picturesque Loch Lochy, the first lake on the Caledonian Canal. In very clear weather, some of the buildings of Inverness, and the high ground of the Black Isle beyond, can be made out, while the rounded outline of Ben Wyvis distinguishes it from the sharper summits of the Ross-shire hills to the N.W. Passing over the wilderness of hills in northern Inverness-shire—among which Mam-Soul and Scour-Ouran are the most prominent—the jagged outlines of the Coolins in Skye may be easily detected rising above the lower heights between. Looking over the head of Loch Eil, the islands of Rum, Eigg, and Muck form a magnificent background to the rather tame slopes of the mainland; and in the clear space of sea between Rum and Skye, a few heights of the Outer Hebrides, distant about ninety miles, may be discerned.

Southward, the dark ridges of the Glencoe hills overtop the nearer bens on the further side of Glen Nevis. Beyond them, the horizon is closed by the long ridge of Ben Cruachan, and the tumbled masses of hills to the eastward of it. To the right of all these, Loch Linnhe stretches away towards the open sea, bounded on the further side by the hills of Morven and Mull, exposing to view the island of Jura with its two conical peaks, Islay almost hid by its sister isle, Colonsay nestling behind

Mull, and the many craggy islets that dot the sea between. On exceptionally clear days, when the horizon is free from all haze, the high ground of N.E. Ireland is seen as a dark line crossing the ocean between Jura and the mainland. Southward from Ben Nevis, the desolate plateau of the Moor of Rannoch, stretching away to the wilds of Perthshire, is easily followed. Beyond it, the most prominent hills are the double peak of Ben More and Stobinian, the elevated ridge of Ben Voirlich, the rather shapeless heights of Ben Lawers and its neighbours, and, standing alone, the prominent cone of Schiehallion is clearly outlined on the sky-line.

In sultry weather, sheet-lightning, or "wild-fire," as it is dubbed locally, was frequently to be seen from the Observatory, and a magnificent spectacle it was. In order to witness the varying effects advantageously, darkness was desirable, and while no opportunities in this respect were lost, it was, truth to tell, not a little eerie to stand at that height viewing the fantasia of the electric fluid. At one time, losing itself behind a bank of cloud, at another flashing on the sky-line, it would re-appear in the zenith, glimmer on the ocean, and betimes illumine some of the nearer valleys, shewing up lakes and corries as distinctly as though it were daylight.

Ere blotting out the summer vision, accompany me in fancy for a turn in one of the passes. Solitude! Unconsciously you give utterance to the exclamation, as, after the precarious descent accomplished on stepping-stones of porphyry, you find yourself in the valley, a mere atom viewing the chaotic residue of a mighty prehistoric upheaval. Boulders and rocks, to the weight of many thousands of tons, lie indiscriminately strewn all over the slopes, and as you brood on eruptions and volcanic action—both of which, in this case, are inept—your gaze turns upwards and rests on the giant Ben, whose immensity you ineffectually endeavour to grasp. Speechless with wonder, you gasp, and are startled at the audibility of your own voice: that death-like stillness—the absence of all sound—these are too much for you, bred to a busy life and boon companions, and to ease your pent-up feelings, you give vent to an unearthly yell, which echoes and resounds from a score of adjacent cliffs. And so is it with many who crave for solitude; their heart's desire is vouchsafed, and the reality almost unhinges their mental equilibrium.

On the midnight watch, in the dead of winter, and in a genuine hurricane, there was a majesty in the elements almost too awful for any soul to experience alone. Thus, for instance, when peering out from the security of the Observatory, the observer felt no qualms, but when "out in it," there was little room for thought. Then occurred a fight with the weather in dead earnest, as with a living antagonist. Again, in the same watch and at the same time of year, with a cloudless sky above,

AN EARLY WINTER COATING.

and the whole country below swathed in spotless snow, and bathed in silvery, silent moonlight, the outlook from the Ben disclosed a terrible grandeur. Viewed either in summer or in winter, at sunrise or sunset, at noon or at midnight, the gigantic panorama reaching to the convergence of the heavens with the earth, diversified by mountains, mists, and lochs below, and fantastic forms of fleecy clouds above, with the ever-changing effects of sunlight and moonlight, the world from Ben Nevis presented one of those impressive scenes to which the artist, man, even at his best, can do but scant justice.

On the approach of spring, how soul-satisfying it was to sit on a sun-exposed slope, lazily dreaming away the time, basking in the sunlight, and enjoying the enthralling vista to the west: the calm expanse of Loch Linnhe, in the distance the ragged peaks of Skye; the green hill-tops elsewhere rising through the soft, blue haze; the peace, and silence, and quiet harmony of everything; no sounds save now and then the croak of a raven, the bark of a fox, or far, far down in the smiling valleys, the plaintive bleat of a lamb, the hoarse call of the red deer, or the subdued murmur of some distant waterfall. Could anything be more soothing, more inspiring, more refreshing both to mind and body? True, there was a qualified charm, when winter had fled, in returning to lower levels; a charm in the varied colouring after the perpetual white glare of the snow; a charm in the old familiar sounds and sights of a country spring-time; a charm, too, in seeing fresh faces of men, and more particularly of women; but when all is said and done, the heart of the mountaineer sooner or later responds to the call of his first love, and not for long is he happy unless solaced by personal communion direct with Nature herself.

CHAPTER XII.

WINTER AND SUMMER PASTIMES.

On Ben Nevis, advantage was taken by the observers of all opportunities for indulgence in sport, and although no very extensive field existed for this purpose, still, being men of inventive faculties, it was surprising what an amount of healthy recreation they improvised, and, needless to say, enjoyment was their's to the full.

During winter, the chief attraction, of course, was tobogganing, and this exhilarating pastime they prosecuted with the greatest gusto, putting to shame the proverbial "greased lightning," as, with a "swish," they rushed down one of the steep declivities which converge on the plateau a few hundred feet below the summit. With lusty shout, and faces all aglow, up and down the steeps they scampered the live-long day, sometimes coming to grief at hazardous points, but never sustaining injuries more serious than a few scratches. Think not that the observations were neglected on these occasions, for, whatever else he might have been, the meteorologist was the soul of punctuality, and the readings, with very few unavoidable exceptions, were always taken sharp at the hour.

To vary the mode of propulsion, it was decided one winter to rig up a sleigh with a sail, and the venture nearly caused the death of a member of the staff. The "boat" was duly floated, and, on her maiden voyage, made an encouraging start near the Observatory, in a steady wind. The skipper, unfortunately, was lacking in knowledge nautical, and foolishly tied the sheet. A sudden gust caught the frail craft and heeled her over on to one runner, skimming along on which she made a bee-line for the brink of a buttress. The man aboard was then in imminent peril, but, as the fates had so decreed, he managed, at the psychological moment, to throw himself from the sledge, which, next instant, went slap over the abyss, leaving him a foot or two from the edge, half paralyzed with fright, and shaking like the leaf of an aspen tree. This was the first and last attempt to establish sailing as an item of amusement there.

When the weather permitted, excursions on snow-shoes and ski were sometimes undertaken, but none of the devotees aloft ever acquired any great measure of proficiency on the somewhat ungainly gear. Not to put too fine a face upon it, it should rather be stated that the antithesis would be nearer the truth.

When cutting capers on ski during a fine day in February, a member, through negligence in strapping it, had the ill-luck to lose one of his "big skates." As chance would have it, it slid over a snow cornice, and stuck in a ledge some twenty or

A TYPICAL WINTER SCENE.

thirty yards down the face of the precipice. Nothing would satisfy the owner but the recovery of his plaything, and, after many misgivings, he was let down by a rope to fulfil his purpose. Securing the ski, he called on those above to haul up, but, horror of horrors! the two who had let him down were unable to pull him back again. In consequence of the rope cutting into the cornice, purchase was considerably lessened, and to make matters worse, the limbs —especially the hands—of the men at the top, became so benumbed with cold, that it was all they could do to retain their hold. A tragic position surely, in all conscience, and it seemed as if nothing short of a miracle could avert a catastrophe. By good luck— or should I say, in consummation of said miracle?— several members of the Scottish Mountaineering Club arrived just then on the summit, and soon disappointed the gravest fears by hoisting up that ski-man, who, like his companions, had suffered untold agonies from frostbite.

Ben Nevis did not escape the ping-pong craze, but the dwellers there had a patent way of their own of playing the game. After erecting a great mound of snow, they squared the mass, which, in an incredibly short time, froze hard, and then, fixing up a temporary screen, proceeded to indulge in the erstwhile popular pastime with the temperature at 17°. The agility required in following the game, kept up the circulation in the players' veins, and the sport was engaged in with as much zest and pleasure amid these unique surroundings as in a West End drawing-room.

When other things failed, or inclemency of the weather intervened, recourse was had to the digging of immense holes in the snow—for exercise to these recluses was deemed as necessary as diurnal rations. Shafts from 10 to 12 feet in depth were, on occasions, sunk, and keen competition existed between the diggers as to who would be first in reaching *terra firma*—if granite agglomerate can be so designated.

During the long winter nights, the most popular pastime within the Observatory was whist, but as the complement of residents rarely exceeded three, one hand, called "dummy," had of necessity to be exposed. One can readily understand, therefore, how welcome a visitor was, if for nothing else than to make up a full whist party.

Several members of the staff at times tried their "'prentice hand" at carving, and a few very passable specimens of the art resulted from their toil. Another employed a whole winter in constructing a model of the Observatory, and this he now cherishes with especial care, as a memento of many happy days spent in peaceful rustication within the original up in cloudland.

Musicians, too, there were, amongst those isolated mountain habitants. The violin, flute, mandolin, accordion, and even the great Highland bagpipes gave forth melody in turn, and on special occasions an impromptu concert would be organized, followed, and brought to a close by a dance, when "hoolochans," jigs, and "hoochs" were very much in evidence. One winter's evening, a Fort-William gentleman treated those above to a gramophone recital by wire. The apparatus was set up, and the records reeled off in front of the telephone at the low level station, while those at the summit, with separate receivers, enjoyed to the full the diverting and novel entertainment.

No more enthusiastic game probably exists in Scotland than that of curling, and its votaries spare no pains or expense when an opportunity arises of engaging in their favourite pastime. Highlanders may be "canny" in some things, but whisper to them that there is playable ice within a day's march, and you will marvel at their agileness and alacrity. Time was when the lake mid-way up Ben Nevis served as a rendezvous for devotees of the "roaring game," and to be a spectator, merely, of the attendant fun, was well worth the exertion expended in reaching the "pond." Stones, brooms, *uisge-a-beatha*, etc., were transferred thither on pony-back, and nothing short of a thaw could damp the ardour of these gamesters. Having prepared the surface, the rollicking sport commenced without any ceremony, and anon the hills would resound with the crack of colliding stones, the ring of the ice, and the inevitable shouts of "Soop, soop, soop!" These were days of unmingled pleasure, and when Nature and duty allowed them, the observers were not loath to hie themselves lakewards, in order to participate in the joviality. The custom of foregathering on Lochan-Meall-an-t-Suidhe has, alas, like the Observatory, now been discontinued; but while the loss in the one case is merely local, in the other it is national.

In summer there was no lack of amusement, but as it was principally ephemeral, and depended to some degree on the visitors, the task of differentiation is one not unattended with difficulty. There was a simple pastime, however,—to wit, rope-quoits, invented by one of the staff, which obtained much vogue during fine weather. The game, which was played on the roof of the Observatory, took the same form as ordinary quoits, the only difference being in the "quoits" themselves, which were composed of pieces of rope spliced into spheres. Frequently, when a bout would be in progress, a clique of tourists would join in, and the excitement and novelty of the sport evoked considerable interest on the part of bystanders. I recollect on one occasion of a German challenging an onlooker to a points game for pence, and the antics, expressions, and exclamations of that alien, made us laugh so much that tears trickled down the cheeks of several who watched the proceedings.

A not uncommon practice with visitors—the staff themselves were addicted to it at times—consisted in the hurling of large boulders down the great cliffs

A ROLLING SEA OF FOG.

and precipices of the Ben. By joint action, stones weighing nearly half a ton would be forced over, and what a din and commotion ensued! Very seldom did a boulder make the clear drop of 2,000 feet, but a short distance on its course, coming in contact with a projecting rock, it would knock off pieces probably larger than itself, and these again, starting others in their downward flight, made the cañons give forth sounds comparable only to a battery of artillery in action. When any ploy of this nature was in the wind, sightseers would place themselves where they could advantageously watch the resultant effects, and, from certain coigns of vantage, it was possible not only to hear the boom and thunder, but to witness the inception, growth, and completion of the avalanche. Some even were known to descend to the bed of the valley, and, from a place of safety there, would gaze with awe on the rumbling tons of débris as it tumbled and broke and roared adown the abyss.

Photography, as may well be imagined, was a hobby which filled in much of the observers' spare time, and Ben Nevis presented a wide range to those with leanings in this direction. Mayhap the enthusiast in quest of unique snap-shots would go as far afield as the half-way station, where, naturally, he would fraternize for a time with those doing duty there. An *al fresco* meal, perhaps a bathe in the lake, would brace him up for the 2,000 feet of climb which lay between him and his "diggins," but he would scorn the path, and make the ascent by some circuitous route, in the hope of securing something fresh to "shot."

In the height of summer, when, for weeks at a stretch, the sun temperature mounted to 130°, it was difficult to shake off an inherent feeling of lassitude. Rambles of exploration were then at a discount—far more satisfying was it to lounge on the Observatory roof, though, even in this, one had to be careful, if respect were to be paid to the nether extremities, for, so hot did the lead become with the sun's rays, that it occasionally proved unbearable to the touch. In company with one's pipe, and a readable book, in a shady nook, fanned by a zephyr's breath, and the whole world at his feet—what more could the heart of man desire?

Ah me! To hark back to those halcyon days and think they are no more, is sadder than I ever thought it should be; but the memory of them, thank God, is spared to me, and in my soul it shall live till the dawn of that fuller life, when the things of sense and time will give place to that more glorious elysian day, when there shall be no more partings.

CHAPTER XIII.

EXTRACTS FROM THE VISITORS' BOOK.

To the average individual, a visitors' book has very few attractions, and, it must be admitted, that the perusal of such records is usually tedious, and not calculated to produce any edifying results. Who, for example, wants to know that Mr. and Mrs. So-and-so stayed at this or that hotel, and were very comfortable; or that the tariff at Blank's Boarding House was exceedingly reasonable, and the attendance excellent? Occasionally, however, one does come across quaint and pawky inscriptions in the visitors' books in remote Highland hostelries, but the ubiquitous tourist is not blessed with the perseverance and patience necessary for the unearthing of these, and so the wise and piquant sayings are lost to posterity.

The Observatory Visitors' Book, which lay at the hotel during the time that houff was open, was a volume of peculiar interest, and a study of its pages helped greatly at times to ward off ennui. Many of the contributions were not without merit, but although a proportion of them bore the stamp of originality, it was not difficult, here and there, to come across flagrant cases of plagiarism. Doubtless, some who scan these lines have, at one time or another, wasted the observers' ink endeavouring to indite a poem or propound a pun on Britain's highest Ben; and if by chance any of their efforts are harshly dished up for public scrutiny in the pages which follow, they must just submit to it as being one of the penalties of fame! Some of the black and white sketches were really mirth-provoking; but when scribblers descended—or ascended, perhaps—to Greek, Latin, French, German, and Italian, or employed the mystic signs of phonography, the unlettered grumbled; and who knows but many literary gems were thus lost to the world! Punsters galore evidently climbed the hill, for, sandwiched in between an otherwise heterogenous collection of nondescript productions, was to be found, on nearly every page, the phrase—"Missed the view, and viewed the mist."

It may be as well, primarily, to dispose of the puns, discarding those of a puny nature, and reproducing only such as would not put the authors to shame. The visitor who inscribed the following line, has a "Note" to the effect that puns are underlined to assist beginners—

"Found the *climb* and *clim*ate produce a most *Ben-Nevis*-ent effect." Probably the following verse, from a punster's point of view, had no equal in the book. The contributor no doubt made the ascent on pony back, and after hours of toil had evidently reached the top only to find it enshrouded in the inevitable mist—

THE ADVENT OF SUMMER.

"In rain's most tyrr'nous reign, our pony's rein we
drew,
Then 'mid mysterious mist it seemed we had missed
the view;
But patient souls, old Sol's bright beams ere long
console,
All Scotia's mountains 'neath our gladdened gaze
unroll."

Tackling the climb with light shoes, as I have already indicated, is a common error, and one probably responsible for the majority of collapses on the hill. 'Twas, in all likelihood, one shod thus who wrote—

"Lost my heel, but saved my sole!"

After refreshing the inner man at the hotel, one broiling day in summer, a wag turned to the book of records, and penned these lines:—

"Oh! Visitors' Book, what strange tales you tell,
Intending, no doubt, to astonish the ages;
Yet tourists, when they reach this hotel,
Feel they must lie somewhere, and lie on your
pages."

Glancing at the comic element, the eye may alight upon such lines as—

"I climbed this hill, I wrote this lay,
And in this book I signed me;
May I come back some happier day
With the girl I left behind me."

Ben Nevis supplies unlimited quantities of literal mountain dew, but the concoction bearing a synonymous appellation is conspicuous by its absence. To discover this only on reaching the summit, must have been galling to seasoned topers, though one is inclined to sympathise with the party who wrote—

> "We've reached the top, upon my soul,
> Without 'Long John' or 'Bogie Roll.'"

"Long John," it may be explained, is the name given to a blend of whisky manufactured in the district. The poet of another company of forlorn mountaineers unburdened his soul thus—

> "We toiled along with saddened hearts—and grief,
> And found—ah, well, just mist and tinned Australian beef."

The mist, as may readily be imagined, came in for a large share of censure; in fact, the majority of those who had the temerity to launch into verse, generally took this as their subject. Here are a few samples—

> "We climbed thy stony sides, oh Ben!
> We groped around thy cloudy head,
> We peered, and jeered, and swore—and then,
> In sheer disgust, we went to bed."

No apology is made to Byron by the author of the following lines—

> "Roll by, thou dense and damp pea-soupy shroud!
> Do we thus reach the highest point in vain?

Roll by ! we say, and leave behind no cloud
Our view to mar ; but, should'st thou still remain,
Mark well the threat—'Never shall we come again.' "

Here is another coming under the same category—

" Befooled, enraged, footsore we trudged,
With aching limbs and freezing sweat ;
Our wasted time and efforts grudged—
We'll never come again, you bet ! "

Mist again is doubtless responsible for the next couplet—

" We came to see the thousand hills,
But have to mourn a thousand ills."

The scribe who engrossed the following verse appears to have been in better spirits, and *a priori*, foregathered with the observers—

" Ben Nevis ascended,
'Long John' ended,
Boots to be mended,
No one offended,
Fiddle and voices blended."

To encourage faint-hearted climbers, was, presumably, the object of the contribution next given. As to the fruitfulness of its mission, history is silent—

" If at first you don't succeed, try again :
Mist and rain you should not heed, try again ;
When the clouds have rolled away,
And the sun holds glorious sway,
Climb the path without delay, come again,
All your labours he'll repay—grand old Ben."

The manufacturer of a famous brand of pills gets a gratuitous, albeit ambiguous, advertisement in the following verse. As the present volume makes no pretensions as an advertising medium, the author has thought well not to disclose the said firm's identity—

"Hurrah! I've climbed the monarch
Of Britain's glorious hills,
And now I'm here, I'll swear it's worth
A ton of ——'s pills."

To those who have resided in, and taken up duty at the Observatory, the contributions next printed will appeal, and the three verses are believed to have been composed and penned by volunteer meteorologists prior to their departure from the summit—

"Farewell, thou sheltering nook,
The highest in the land:
So thou in my esteem and backward look,
Highest shall ever stand."

"Around, oh! what a glorious spread,
With hills and glens the eye is fed;
But oh! 'the Ben,' when all is said—
The mighty Ben for me."

"Ben of peaks the clouds that sever,
Oft thy steeps have wearied me;
Must I leave thy shade for ever?
Then, farewell—farewell to thee."

TOBOGGANING.

Some good practical advice, not only to climbers of the Ben, but to the human race in general, is offered in these lines—

> " Never despair, or the burden may sink you;
> Providence wisely has mingled the cup,
> And 'mid the cares of life, ever bethink you,
> The watchword must aye be, 'Never give up.' "

The poetic flame must have burned in the heart of he who wrote—

> " The sombre hills, the sparkling lochs,
> The burn-scoured glens with birken tree;
> The trailing mists, the sunlit clouds,
> Have filled my breast with ecstasy."

Talking of clouds reminds me of a laughable incident connected with the early days of meteorology on Ben Nevis. One particular evening the weather report to the newspapers contained, *inter alia*, the words—" Sky covered with cirrus clouds." Judge of the superintendent's consternation next day when he read, " Sky covered with circus clowns."

Several of the entries in prose are also very amusing. Glancing over the " Remarks " column at random, such expressions meet the eye as:— " Have never been nearer Heaven," " My kingdom for a horse," " God made it, let it pass," " Leaves a good impression behind," " Had doubts about there being any top," " O for a wringing machine," " Away from the madding crowd," " Why left I my hame?" "I have now discovered the mist factory of Scotland," "Oh, noble Ben, oh, noble we, who sweat ourselves thy top to see," "Pilgrims on the world's highway," "Broke the record, the Sabbath, and nearly broke myself," "There is toil on the steeps, on the summit repose," "Just alive when I arrived," "One dose is enough," etc.

Many of the visitors took a special delight in perusing that volume of impressions, and were in the habit, when not detected, of adhibiting addendums to contributions inscribed by previous climbers. To shew what is meant, a few examples are reproduced here. The phrases within parenthesis are those added—

"O Caledonia, stern and wild."
(Wet-nurse for a poetic child.)

"My end shall be water."
(No spirits here.)

"Every valley shall be exalted and every mountain shall be brought low."
(May the time soon come.)

"The top at last—seventy-five years of age."
(Cheer up! you'll soon be dead.)

"Ascent easy; don't know what the Highlanders have to blow about."

(Not being out of puff, they don't need to blow.)
(Met Critic No. 2 going down, blowing like a porpoise.)

"Mist—a fine view."
(Musty joke.)
(It's no wonder—snow wonder. See!")

The sentiment expressed in the last extract with which I will trouble my readers, permeates the entire book, and I only select the one which follows on account of its graceful setting. "No pen," wrote the enthusiast, "can adequately describe the sublimity of the scene, nor tongue give expression to the awe-inspiring grandeur produced by the wonderful works of Nature."

CHAPTER XIV.

SOME ENTRIES FROM THE LOG.

An examination of the Observatory Log-book is, in the highest sense, a congenial task, inasmuch as its pages throw light upon many subjects of interest not otherwise disclosed by the somewhat dry meteorological records. Instead of attempting to compress the entries into a series of running paragraphs, I have thought it better to reproduce these just as they are stated, and in this garb they will doubtless appeal with more force to the reading public.

1883.

Nov. 23. House-fly to-day seen crawling inside window of living room.

28. A weasel to-day came and looked in at window about 14h. 30m.

Dec. 11. At 18h., the doors were so blocked with drift that no outside observations could be taken. One was made with considerable difficulty at 19h., and then the door was closed for the night.

1884.

Jan. 6. Everything covered with ice, either solid or crystalline. At 4h., thermometer box

PING-PONG ON A SNOW BLOCK.

was frozen so fast that it needed chisel to open it. At 10h., some snow-flakes about 1-inch diameter were observed.

Jan. 13. A small rat found drowned in pail in coal cellar.

21. Very stormy in early morning and evening. At 19h., 20h., and midnight, Mr. Omond and Mr. Rankin took observations roped together. At 19h., a bright light was seen along the edge of the cliff, forming a white glare about 10 feet broad, and lighting up the whole top of the hill. May have been due to snow blowing over and getting illuminated from below.

26. As the drift prevented the reading of the thermometers at 13h., the observers again went out tied together, but found it impossible to go further than the snow porch with safety; at 16h., they got as far as the box, but could not see the instruments in consequence of the drift blowing up in their faces.

Feb. 7. About noon, two birds, thought to be hawks, were seen flying backwards and forwards over the ridge to N.E. of Ben.

9. A white weasel seen at one of the bedroom windows to-day.

Feb. 16. Owing to storm, every observation taken by two observers roped together. At 20h., as soon as Mr. Omond went outside door of snow porch, he was lifted off his feet and blown back against Mr. Rankin, who was knocked over.

Mar. 1. Snow porch at outer door is now about 33 feet long, with a rise of about 12 feet in its floor.

Dec. 23. To-day the feathery snow crystals, forming on exposed surfaces, were grey-brown in colour.

1885.

Jan. 11. At 9h., rain-guage blown away—probably over cliff. The snow-drift blown up the cliff near the Plateau of Storms was seen to rise from 100 to 200 feet in the air, and pass away to leeward in clouds, resembling ordinary small scud-clouds.

Feb. 21. At 16h., the note-book for the observations was torn in two by the wind and blown away. After 17h., no temperature readings were taken, as the lamps could not be kept alight, nor could the observers stand against the gale. At 22h., the glass in one of the tower windows was broken by a flying lump of ice.

1886.

Jan. 1. Robinson tied up at 23h. 10m.

This, of course, refers to the Robinson anemometer, which, under certain conditions, had to be stopped and secured—and thereby hangs a tale. A cook, also named Robinson, soon after his arrival at the summit, developed meteorological instincts, and always scanned the sheets when an observer completed his rounds. Some friction had arisen between him and one of the scientists, and existing relations were not bettered when the culinary man by chance happened upon the entry last given. Going up to the author of it—who, by the way, was the individual with whom he was at variance—he demanded to know what was meant by such an "imprecation" on his character. At this the observer burst into a fit of uncontrollable laughter, but latterly an explanation was tendered, after which a reconciliation took place, and the trio thenceforward lived in amity.

Mar. 27. At 13h., a snow-bunting was seen on top.

Apr. 21. Meteor as bright as first mag. star seen at 23h. 5m. It went from Pole Star towards W.N.W. horizon, and left a train about 10° long at a height of about 50°, which remained visible for a second.

Nov. 11. At 8h., the shadow of Ben Nevis was seen on low clouds to N.W., and with it a badly-defined glory.

1887.

May 27. To-day, two horses with stores reached the Observatory for the first time this year, coming right up to the door over the snow.

Sept. 16. Between 3h. and 4h. this morning, the Zodiacal Light was seen as a steel-grey glimmer extending from eastern horizon through the constellations Leo and Cancer, to the border of Gemini. At the same time, a light, supposed to have been the glare of the Carron Ironworks, was observed on the S.S.E. horizon.

Oct. 29. At 1h. 5m., St. Elmo's Fire was seen in jets 3 to 4 inches long on every point on the top of the tower, and on the top of the kitchen chimney. Owing to the number of jets on each cup of the anemometer, this instrument was quite ablaze. On the kitchen chimney, the jets on the top of the cowl were vertical, and those on the lower edge of same horizontal. While standing on office roof watching the display, the observer felt an electric sensation at his temples, and the second assistant observed that his companion's hair was glowing. At 1h. 15m., the accompanying hissing noise ceased, and the fire vanished from every point the same instant.

THE HALF-WAY STATION AND LAKE—TEA *al fresco.*

1888.

Jan. 2. The tracks of a hare were seen at thermometer box at 8h.

4. At 2h. there was no drift, but junks of ice and hard snow were flying about. At the same hour, the observing book and sheets were blown away, and it was found impossible to carry rain-guage and manage lamp.

Sept. 19. Very fine sunrise. Just before the sun rose, long pink streamers were seen diverging from E. horizon, passing overhead, and converging to W., where a rosy belt topped the earth shadow.

1889.

Oct. 1. At 2h., a faint aurora and the Zodiacal Light were seen. The latter was best seen at 2h. 45m., its southern boundary being well defined. The Zodiacal Light, the "Milky May," and the horizon formed the three sides of a triangle, the enclosed sky being markedly darker than its surroundings.

1890.

Jan. 5. At 19h. 25m., a spark or flash from the office stove struck Mr. Millar, who was sitting at the desk in office. The shock passed from the neck downwards, but did him no injury.

Jan. 29. The snow at the first gorge was measured this afternoon, and was found to overhang the edge by about 17 feet. The drift of snow lying between the first gorge and the hotel is about 10 feet deep.

Feb. 9. At 8h. 30m., two ravens, and at 7h. 45m., snow-bunting seen. Mock sun seen at 9h. 5m.

20. The crystals on the anemometer were to-day fully 7 feet long.

May 6. A rat was seen at 2h.

24. All last night, the Northern sky was bright with sunset or sunrise colours. At 1h., these were red, yellow, pale green, and blue.

1891.

Jan. 31. A severe gale from the south-east sprang up in early morning, reaching the strength of 11 at 7h. At that hour the observations were 10m. late, owing to the observer's lamp having been blown out several times.

April 13. Beautiful snow crystals falling in the afternoon; not in any measurable quantity, only a few now and again; are perfect hexagons, some in star and others in glassy disc shape. Most had small motes of dry snow attached to them, and these motes were always on one side only.

Sept. 15. A large weasel was caught in a trap at 18h. 20m.

1892.

Jan. 12. There are now wreaths of snow about 15 feet deep lying between hotel and first gorge.

May 18. A large rat was seen on the roof at 1h.

Sept. 21. A magnificent aurora seen from 20h. till midnight. There were three distinct arches from W. and S. to N.E., with shifting streamers shooting to the zenith.

Oct. 21. A sudden and severe gale carried off an arm of anemometer and a rain-guage.

Dec. 3. About 10h. 30m., the telegraph instrument indicated several flashes of lightning, the last of which short-circuited the line by breaking through the protector.

1895.

July 25. Mr. Rankin and two others descended the Tower Ridge, crossing "The Gap," with a couple of cameras. Several photos were taken from the vicinity of the cairn.

Dec. 10. At noon, two ravens were hovering over the summit. Mr. Neil saw a small mammal (species unknown) in the visitors' room; he described it as being about 4 inches in length.

Dec. 11. A fox was seen at the Observatory door at 19h.

20. In returning to the Observatory to-day, Mr. Bruce observed tracks of a fox, as well as the spoor of weasels and hare. On the snow, about 2,200 feet, he picked up a number of insects, large numbers of which were crawling on the hard frozen surface. The track, apparently of a fox, was noticed on the Observatory roof, and near one of the windows.

1899.

May 9. A religious service was held on the roof of the Observatory to-day, by the Rev. John M'Neill, of London. Upwards of forty visitors attended.

Oct. 15. This morning, the two observers, accompanied by Mr. Fraser, Glen Nevis, made a thorough search for the dog which for some time has been heard howling down the precipices, with the result that the animal was found on a ledge of rock. It had been at least six days there; but, though greatly emaciated, soon recovered on being taken to the Observatory.

1900.

Mar. 23. To-day, Mr. D. Cameron, along with Mr. Hugh Mackenzie, roadman, left the Low Level Observatory for the Ben.

FOG IN VALLEYS.

The former became done up near the Crystal Well, and the cook, who went to meet the party, had to assist in getting Cameron back to Fort-William again.

June 12. During a severe thunderstorm, discharges and explosions occurred in the office stove, and, on account of the lightning damaging the cable, no report could be sent to the newspapers to-night.

There are many other items of interest in the Log-book, but, to obviate any tediousness, I refrain from quoting further, and must hasten on towards completing what you, dear reader, may already have christened, "these somewhat disconnected reminiscences."

CHAPTER XV.

FAUNA AND FLORA.

WHILE it cannot be said that the fauna indigenous to Ben Nevis is of a diversified order, still, a more than common interest attaches to it, in respect not only of haunt, but also in the matter of species. The largest specimen of *feræ naturæ*, other than the red deer, known to exist on the mountain, is the fox, although it was only during the severity of winter that reynard was seen on the summit. Their tracks were noticed time and again on the Observatory roof, and once during night time, when the observer was going his rounds, he descried a vixen peering in at the kitchen door. The natural timidity of the animal is proverbial, and the one in question must have been emboldened by the pangs of hunger ere venturing so near to the abode of man. Discretion, nevertheless, outlived the pain of knawing vitals, for, whenever the beast caught sight of the observer, it scampered off.

That they bred in the rugged fastnesses of the lower reaches there could be no doubt, and so inimical to lamb life did they become in spring, that organised raids for their extirpation were periodically carried out by the farmers whose flocks pastured on the hill. Think not that a den was easy of location—the opposite being rather the case, but when "Tally ho!" rent the air, no quarter was granted to those enemies of the fold.

Near the head of the glen bearing the same name as the mountain, the wild cat still stalks and hunts, but this species is annually deteriorating, and a few years probably will see its extinction in these parts. It is much sought after by taxidermists, more particularly for its beautifully marked skin; but the hunter must have all his wits about him, as the feline is much given to treachery and is notorious for ferocity.

Herds of goats are occasionally to be seen on the slopes of Ben Nevis, and to watch these sure-footed animals gracefully skipping from ledge to ledge on an almost perpendicular ridge, suggests to one the idea of chamois stalking on the Alps. These goats, be it remembered, are not of the domestic order, but live exclusively on the steeps, where they roam and multiply in a wild state.

Stoats and weasels frequent the summit in considerable numbers, and a peculiarity about these denizens is, that while their fur is of a brownish tint in summer, it assumes in winter a colour almost akin to snow in whiteness. At times they even ventured inside the Observatory, and at least one specimen was captured and sent to headquarters as a rarity.

The mouse tribe was well represented at the meteorological station, but how these mammals gained a footing at such an altitude is a question not easy answered even by those best competent to judge. Voles, shrews, and rats were caught at different times, and excited the keenest curiosity, particularly as regards their teeth, which, in front, were generally found to be extremely prominent and protruding slightly. The chief enemy of those rodents was the Observatory cat, who, immediately on making a capture, came purring up to one of the staff and laid the prize as a gift at his feet. Poor tabby succumbed to senile decay, having, with the exception of about a month, whiled away the alleged nine lives of fiction, at the highest house in the land.

As pets, the observers could boast of a hare, a tortoise, and a frog. The hare, which was brought up a leveret from below, got very tame, and sidled about from office to kitchen quite unconcernedly, and its counterpart in that hypothetical race, known to us all since school days, was not behind in the matter of sociability. Froggy sat in one position on a flower-pot for nearly a year, during which time he was never known to eat, although repeatedly tempted with flies and other "paddockian" delicacies. In time, too, he passed the way of all flesh, after a truly remarkable fast.

Two unaccountable instances of dogs finding their way to the summit may be set down here. One morning, when taking the outside readings, the observer's attention was attracted by a whining noise emanating from the visitors' room. On going

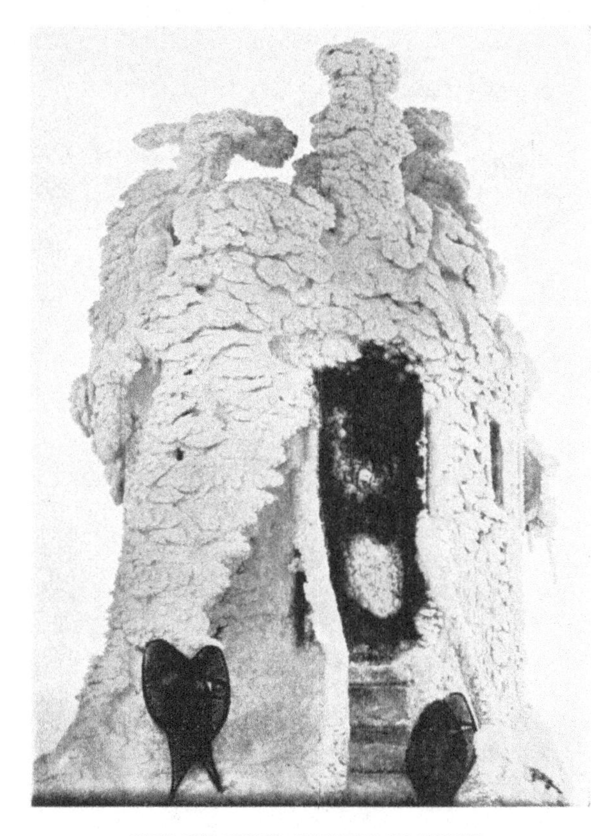

HOW THE TOWER APPEARED IN WINTER.

thither, he was nonplussed for the instant on discovering, in a box of straw, a bitch with a litter of six pups. How the animal sought out such a place for depositing her offspring, or whence she came, will probably never be cleared up. The pups being of the *highest* pedigree, were not long in finding homes. The other case is that of a dog which, for many nights, was heard howling somewhere down the precipices. Numerous attempts were made to sight the animal, but without avail, and at length the two observers, aided by a visitor, went down the rock-face roped together, and found the dog on a jutting ledge of rock. When discovered, it was very thin and unable to stand, but on being taken to the Observatory and fed, it soon revived. It must have been about a week in the corry, but the mystery of its presence there has never been satisfactorily explained. Ownership was traced to a Fort-William citizen, and to him the dog was shortly afterwards sent.

Bird life on the hill was more marked by its paucity than redundance, but the subject, notwithstanding, afforded much scope for study. The only species which actually took up residence on the mountain-top was the snow-bunting—a most interesting little bird, always to be seen flitting about the Observatory. Though hatching in the neighbourhood of the cliffs, no single instance of a nest having been discovered is on record. Of a greyish-brown colour in summer, their plumage in winter became almost pure white, making it difficult to distinguish them in the snow.

Hawks and falcons were sometimes seen hovering over the summit, and they must have preyed upon the buntings, for it was no uncommon thing to find the feathers and other remains of the latter in places which left little doubt as to the marauder. Ptarmigan existed and nested at about a height of 3,000 feet, and were occasionally observed flying across the hill-top. This, too, is another specimen of the bird family which has the power of changing the colour of its feathers with the changing seasons. At rare intervals, a glimpse might be caught of the golden eagle soaring at a height perhaps double that of the Ben, while ravens — the croak of which could be heard for miles — were often to be detected, especially during the summer months.

At such an altitude, one would have expected a comparative absence of insect life, but such was not the case. The common house-fly made his presence felt within the Observatory; and without, on a summer's eve, that hardy annual commonly referred to as the "midge," was wafted upwards on air waves, and would not be denied his quota of blood. Butterflies of varied hue and size helped to swell the throng of summer visitants; and, on the vanishing snow-wreaths, one could spend a very profitable hour or two inspecting the countless myriads of coleoptera. The existence and number of these insects on the snow patches was, to say the least of it, incongruous—that is, from the point of view of a novice, though, to a naturalist, the seeming phenomenon might be simple enough of explanation.

There is no great variety of flora on Ben Nevis, a fact due, doubtless, to the meagreness of vegetation on the upper shoulder; but what there is possesses a peculiar charm for the botanist, alike on the score of rarity and beauty. Near the Observatory well, which is only about 65 feet below the Ordnance Survey cairn, and between the lichen-covered blocks of porphyry, are small accumulations of peaty humus, moistened by the drippings from the contiguous springs, where several plants of the order *Saxifraga stellaris* are to be found. This is the highest point in the British Isles where the flora named is to be seen, exceeding that from the summit of Ben Macdhui, 4,296 feet, which is recorded by the late Dr. Dickie. In the vicinity, two other flowering plants thrive, viz., *Deschampsia coespitosa*, and *Carex rariflora*, and these never exhibited any sickly tendency, but on the contrary were always fresh and strong. This points to the belief that the absence of phanerogamic growth on the summit is not altogether due to altitude nor low temperature, but more to the scarcity of suitable accumulations of soil wherein plants could take hold.

Numerous fern species rear their green crests amid the crevices, the rarest being that known as "Parsley," which is, to all intents and purposes, an exact replica of the vegetable bearing a synonymous cognomen. This cryptogamic specimen requires practically no soil to ensure growth, and flourishes in humid recesses all over the mountain, having been found as high up as 4,200 feet. During the season, tourists were in the habit of uprooting and carrying off all obtainable clumps, but the plant invariably drooped and died at sea level, being presumably dependent for existence on altitude, and the conditions obtaining on its native Ben.

There is a wealth and dissimilitude of mosses to be met with on the slopes, though it requires the eye and vigilance of the enthusiast to unearth the more cherished varieties. In colour, these herbaceous musci vary from dark brown to almost pure white, and from all shades of green to the deepest magenta.

Intermixed with the heath and brackens which bedeck the lower dells, the much-prized white heather blooms prolifically, and lovers of the rare spray, who diligently search for it here, seldom return unrewarded. The blaeberry and the cranberry grow to profusion in some of the defiles, and tempt the palate of tired wayfarers ; while the delicious aroma of bog-myrtle and wild thyme helps materially to lighten the burden inseparable from climbing.

A CORNER OF THE OBSERVATORY INTERIOR.

CHAPTER XVI.

TROUBLES—PHYSICAL AND OTHERWISE.

As beneficiaries under the universal disposition of ills to which the flesh is heir, the observers had no option but to accept their *jus relictae*, and this they did with the best grace possible. Their mode of living, coupled with the purity of atmosphere, and comparative absence of all germinal taint, secured for them an existence, if not entirely void of disease, at least unmarred by the affections common to dwellers in the plains.

The most serious ailment contracted by any member of the staff was in the case of one of the cooks, who was laid aside for some time with a pretty severe attack of rheumatic fever. Indisposition has its drawbacks at all times, but when the patient is bedridden a mile up in the clouds at midwinter—as in the instance cited—difficulties are multiplied. Twice had a medical man to ascend the mountain at great personal risk, and the danger was not lessened when, on the subject's convalescence, he was strapped on to the back of a pony, swathed in blankets, and by this means conveyed over the great snow-banks to a more salubrious clime.

The very nature of their occupation, viewed in relation to the fluctuation of temperature, would naturally lead any one to the conclusion that these weather men would be often subject to catarrhs and kindred throat inflammations, but experience shewed quite a different result. The subsequent residence at sea-level, however, rendered the members of the staff peculiarly liable to an affection which they accustomed themselves to designate as the "Ben Nevis cold."

"This closely resembled," to quote from a note on the subject in the *British Medical Journal*, by A. Cameron Miller, M.D., F.S.A., Scot., "what we are in the habit of recognising as an influenzal catarrh. The attack generally commenced in about forty-eight hours after the patient had taken up residence at the low level. First, feelings of shiveriness were experienced, followed by *malaise* and disinclination to exertion of any kind, stuffiness of the nasal passages, sore throat, hoarseness, sometimes cough, coryza, and occasionally tenderness of the conjunctiva and lachrymation. The temperature did not shew much disturbance; usually it did not rise above 100° in the evening, although the symptoms became aggravated in the afternoon. Gradually, for a period varying from ten days to three weeks, the patient got rid of the objective symptoms of the attack. During the whole period, he felt uncomfortable and disinclined for work, and that condition persisted for some time after the coryza had disappeared. One attack did not seem to have the least protective influence, as recurrences took place in the same individual time after time, but intermissions did occur. Post-primary attacks might be as severe as a primary attack, and an attack following upon one month's residence on the summit might be as severe as an attack following the usual three months' residence. The period of the year apparently exerted no special influence, summer attacks being quite as severe as those noted in winter. Certain individuals of the staff appeared to be more susceptible than others; and a member suffering from an ordinary cold at the low level, on mixing with his colleagues resident on the summit, almost surely infected them with catarrh, which, however, seldom endured longer than twenty-four hours. The history of the interesting condition here sketched, points with little doubt to germ influences in the lower atmosphere. At the top of the mountain, organisms do not exist, or, if they do, exist only in inoccuous numbers. At sea level they relatively thrive, and, seizing upon the 'virgin soil' of a renewed and susceptible mucous surface, they set up the irritative and mildly toxic phenomena described."

In a separate paper read before the British Medical Association, Dr. Miller pointed out that, under certain circumstances, advantage might be derived from high level residence in the treatment of tuberculous conditions, and advocated the establishment of a sanatorium on the summit of Ben Nevis.

Several of the staff were great martyrs to toothache, and diversified were the remedies resorted to, in order to obtain relief from the gnawing pain. From the tool-box to the medicine-chest, first a pair of plyers then a cocaine lotion, followed alternately by a mustard plaster, a concoction of salt and brandy, a mouthful of snow, an Irish jig, and lastly, if all these failed, a strenuous endeavour was made to "kill the nerve" by prodding the offending molar with a needle. This "hell of all diseases"—to borrow the nomenclature of Scotia's bard—is verily a caution when the sufferer is located outwith the hail of a dentist, and none understood the truth of this better than those resident on Ben Nevis. Indeed, in one or two acute cases, nothing would satisfy but extraction, and the journey up and down the hill was gladly undertaken in an afternoon, rather than endure the excruciating agony longer.

Tonsorially, the meteorologists underwent some minor troubles, but these they treated with stoical indifference. A relief man from below, during his first week of occupation, made commendable endeavour to keep up the practice of shaving, but the second application of the "razor-saw" was usually the last during his stay on the summit. Mutual accommodation once prevailed in the matter of hair-cutting, but a laudable effort on the part of a new cook, resulting in the reproduction of a guy, put a stop to this custom. In these circumstances, the hirsute appendage and the top thatch were allowed to accumulate at will, and oft-times, especially in winter, these disciples of science presented more the appearance of wild men of the woods than specimens of cultured humanity.

Troubles there were, too, in the procuration and retention of suitable cooks. Unlike the observers, the master of the culinary department had to reside constantly on the hill-top, and in course of time this continued isolation became irksome, so that when a good man had been secured, the difficulty was to get him to remain. Some of the chefs were bright samples of the *genus homo*, and were not above assuming airs when opportunities occurred. I myself have heard one of the kidney inform a visitor, whom he was showing over the Observatory, that he was the first assistant, and this fellow, I candidly believe, could not tell the respective uses of the barometer and anemometer. Another was so scared at the lightning during a severe thunderstorm, that, instead of retiring for the night, he chummed with the watchman and accompanied him every hour when he went out to take the readings. One such night was enough for that cook, and he gave notice the next day.

Talk of stage-fright! Faugh! it fades into oblivion when pitted against the mountain scare. A temporary observer one season expressed a strong desire to take a turn at the night-shift, and a favourable opportunity arising, his wish was gratified. The night was an ideal one, towards the latter end of summer, with a full moon, cloudless sky, and a dead calm—nothing, in fact, awanting to satisfy the most finical. The volunteer took up the watch accordingly, but after taking the observations at 2 A.M., he had to arouse the man who regularly did the work, as he found it impossible to remain up alone amid such surroundings. The measured tick-tack of the clock within, and the death-like stillness without, where all was as the tomb, had the effect of unnerving him, and he told his companion afterwards that, had he been left another hour alone, he would literally have gone mad.

Another source of frequent trouble to the staff, was the pertinacity of some of the visitors, who, at all hours of the night and day, would find their way into the Observatory. In the cold of the "sma' hours" they craved for warmth; at mid-day, water was chiefly in demand. During the night watches, when all were abed except the man on duty, it was roiling to be besieged by a noisy party begging for victuals, and, although it seemed churlish to turn them out, what alternative was there? The institution was not erected *pro bono publico*, and if the lavish hand had not been stayed, the visitors would have fared high, and the meteorologists would have starved.

Hospitality, nevertheless, was extended when circumstances warranted the same, and while some may have considered the recluses hard-hearted, there were others, in number not a few, who expressed themselves in different terms. It would require an adamantine heart to deny an aged person after such a climb, nor was it done; and the occasional pitiable plights of the fair sex, appealed to the man at the wheel. From the records, I see that at least three octogenarians made the ascent, the oldest being eighty-three; and, on the other hand, a baby girl, blessed with only three months' existence, was carried to the top by her mother. Item—Which was the more foolish, that old man, or the maternal relative?

The even tenor of meteorological life at the summit was interrupted by the Boer war, and while this may not, properly speaking, constitute a trouble, no harm can accrue from the notification of the fact at this stage. Six men in all from the Observatory staff served their country in South Africa, and, as this is exclusive of the cooks—most of whom were ex-army or reserve men—no one, surely, can impute disloyalty to Ben Nevis. Of these, two served in the Cameron Highlanders, one in Lovat's Scouts (killed), one in Scottish Horse, one in 12th Lancers (wounded), and one in a cycling corps. Under the Southern Cross they wrought well and they fought well, repeating Scottish history, and making it; but, above all, maintaining and upholding the fair fame of the Highlands—that rugged western realm from which hath sprung the flower of Empire's warriors, that region of mountain and flood—"*Tir nam beann nan glean 's nan gaisgach!*"

So much for the staff's troubles, which, for the time being at any rate, are at an end in so far as they relate to residence on the summit of Ben Nevis. Pessimists allege that the reign of meteorology on the mountain has received its death-blow; but there are not awanting those who hold more hopeful views. Personally, I always maintain that the end is not yet.—*Q. E. D.*

CHAPTER XVII.

MOUNTAINEERS.

ONE bracing afternoon in winter, not long after the Observatory had been opened, a member of the staff who had gone out for a walk on the snow, was more than amazed on looking over a precipice to see, about 1,500 feet down, two dark objects laboriously scaling the ice-covered and snow-clad face of the declivity. That human beings should attempt to reach the top by such an access never crossed his mind, but this in reality was what was happening, as a closer inspection shewed him. Roped together, and cutting foot and hand-holds with their ice-axes, slowly but surely those two venturesome climbers made the top by a route which had never, within mortal ken, been previously attempted. To say that the observers were astonished, would be but meagrely to express their feelings, for, had they not, like the Russians and Port Arthur, deemed these gigantic precipices impregnable to frontal attack. Still, there were they communing with the assaulters, and no alternative but capitulation remained, the sequel being a retiral to the domicile, where a right royal repast was spread. These two climbers, along with several others, constituted the nucleus of what is now known as the Scottish Mountaineering Club, and of one of their Easter meets on the Ben I would shortly write.

Amongst the Club's members are professors of various universities, clergymen, doctors, advocates, artists, engineers, and business men, who find the highest form of relaxation and enjoyment in pitting themselves against the forces of Nature, and grappling with physical difficulties which call for the exercise of pluck, endurance, and resource. Theirs is a strange pastime, but members aver that, by thus securing for a time absolute mental rest, they are enabled to return to intellectual labours with renewed vigour. The mountaineering of the Club is not of the paper or "sympathy" description. All affiliated to it can shew a considerable record of ascents made in Scotland under winter conditions—that is, under ice and snow, summer climbs being relegated to the category of hill-climbing. Many are also members of the Alpine Club, and have been engaged in historic explorations and climbing on the greatest mountain ranges of the world.

For the meet, to which I have referred, over thirty members assembled at Fort-William, and directed their attention to Ben Nevis and the surrounding summits. Many came from the northern counties of England, some as far as London and further south, and the first few days were employed in scaling the rock ridges and snow gullies embraced within the circle of what is known as the northern precipice of the mighty Ben. The misty, cold weather which prevailed on the occasion, far from deterring, only added to the zest with which these mountaineers tackled their work. They did not seem to revel in the view from the summit so much as in the enjoyment of surmounting the difficulties of the tasks which they imposed upon themselves. The tourists who visit Ben Nevis in the later and wetter summer months, climb the hill by the well-known path, up which, the mountaineers say, any man with vigour enough may push a wheel-barrow. On the other side of the mountain—the side which tourists seldom ever see, or only see in limited glimpses from the summit—there are at least ten different routes of ascent. Some are gullies filled with snow, which does not disappear even during the hottest August; the others are over or skirting rock ridges of the most precipitous contour, at places almost perpendicular, and, near the summit, thickly plastered with snow or glazed with ice. They are some 2,000 feet in height, and afford in places but the most meagre hand-hold, which is made much more precarious by the cracks having to be cleared of ice before the climbers' fingers or boot toes can be inserted.

On the principal day of the meet, no less than six parties, numbering in all some twenty-eight climbers, made the ascent by these irregular and difficult routes. The temperature was about 3 deg. under freezing point at the time these mountaineers were adhering like so many limpets to the ice-covered rocks and snow faces. One party of three were on the rocks of what is known as the Tower Ridge for no less than nine hours, being absent from Fort-William in all fifteen hours. The upper part of the ascent was made very arduous by the frozen condition of the ridge, which, coupled with the prevalence of fog, made the climb one not unattended by a certain amount of danger. Another party of five made the ascent by the North-East Buttress, at one time considered inaccessible, and only climbed once previously, but since, frequently undertaken and successfully accomplished.

Other members found their diversion in ascending snow-slopes, sometimes hemmed in on either side by walls of black rock hundreds of feet high—the slope of the snow appearing to the uninitiated only a degree less than the perpendicular. It is amazing with what alacrity Alpine men will tackle such a snow-slope, and if an overhanging cornice bars their way, they think nothing of tunnelling a passage up through it—a feat of some magnitude, seeing that perhaps a depth of 20 feet of snow has to be penetrated with no better accoutrement than an ice-axe.

Of the exhilaration and health-giving qualities of this exercise, no one who has had the good fortune to be present at the evening dinners of the club during a meet can have any doubt. The ruddy faces of the mountaineers, which have been exposed to the biting blast for hours on the mountain, the cheery voices, the ringing laughter, and the rapidity with which the viands disappear, speak volumes.

The entrance hall of the hotel reminded one more of such Alpine climbing centres as Zermatt or Grindelwald than a West Highland tourist resort. The umbrella stand was innocent of a single gamp, but was filled instead with a few dozen ice-axes, while coils of mountaineering rope depended from the hat-pegs.

Till the inauguration of the Scottish Mountaineering Club, hardly any one knew how attractive the Scottish mountains were in winter and the early months of spring, while their summits were snow-covered and their corries draped with icicles, and heavily corniced with great wreaths of frozen snow. Some who have seen photographs taken during the spring months, consider they represent views in Switzerland among the higher Alps, having no idea that such magnificent ice and snow scenery is within half-a-dozen hours by railway of the large centres of population.

Talking of railways, recalls to my mind that in this respect Ben Nevis has come in for a fair share of attention, and the time, moreover, may not be far distant when the present mode of ascent by path will be superseded by the modern saloon carriage. Several projects with this end in view have been mooted; nay, more, the mountain has been surveyed, plans prepared, and the route actually staked out in part. Of the scheme I give a few brief details.

The proposal was that the line should start from the base of the hill, in juxtaposition to the West

Highland Railway, follow the old pony track up to the lake, and then, striking off to the east, wind round the upper shoulder of the mountain to the summit. Its total length was stated at four miles, with a carriage road at the lower end of about a mile. As planned, the permanent way would consist of two outer rails and a centre rack rail, laid on sleepers securely fixed together by longitudinal runners so as to form a frame, which would be fixed to masonry foundations at regular intervals. The maximum gradient would be 1 in 2·62 for a length of 600 yards. The undertaking was estimated to cost between £25,000 and £30,000, and it was computed that a remunerative income would be derived, principally from tourist traffic, but the promoters had in view the inauguration of special excursions from the large cities of Scotland and the north of England.

The proposition was to charge as on the Righi, viz., one shilling per mile, single journey up or down, and fare and half return. On the assumption that 50 per cent. of the travelling population made the ascent, the figures, according to the financial advisers of those responsible for the scheme, would work out something like this:—Say 3,750 single tickets at 4s. 9d., and 3,750 return tickets at 7s., equivalent to an income of £2,190, which, after deducting 2½ per cent. for working expenses, would leave for shareholders 6 per cent. on the cost of construction. The erection of an hotel, with suitable sleeping accommodation, on the summit, was also in contemplation, and in such an event, it would doubtless be well patronised.

The proposed railway would not entail much banking or cutting, and practically no fencing would be necessary; but in assuming that the proprietors interested would give the land required, without the necessity of applying to Parliament for powers, the promoters took too much upon themselves, as the late Lord Abinger—through part of whose ground the line would pass—ordained his trustees to oppose any such scheme. Finally, the prospectus notified that the mechanical arrangements would be similar to those which had stood the test for years on the Continental mountain railways, and the belief was entertained that no difficulty would be experienced in satisfying the Board of Trade as to the safety of the line.

Whether or no the privacy of Ben Nevis will ever be disturbed by the shriek and rumble of a railway engine, true it is that the receipts of any company formed for such a purpose will not be materially increased by members of the different mountaineering clubs. On the other hand, there may be good ground for the assertion that a large proportion of the visitors would avail themselves of such a means of reaching the top; but the sentimental side of the question must not be overlooked, nor the crucial one of precedent, under both of which heads much opposition may undoubtedly be looked for.

CHAPTER XVIII.

THE CLOSING OF THE OBSERVATORY.

THE maintenance of the two Observatories, at Fort-William and on the summit of Ben Nevis, involved an average yearly expenditure of close on £1,000. Of this sum, £350 was supplied by the Meteorological Council out of its annual grant of £15,300 for meteorological purposes, the remaining, and larger, portion of the amount having been made up from money collected for the purpose from various private sources. The pecuniary burden thus imposed upon the Directors, and which has weighed more or less for twenty years, at last became so onerous that, in June, 1902, they were compelled to intimate that, as they could no longer raise the necessary funds, the Observatories would have to be closed in October following.

The publication of this resolution led to questions being raised in Parliament on the subject, and the appointment of a Treasury Committee to consider the whole question of the annual grant made to the Meteorological Council. As there was reason to hope that the enquiries of the Committee might lead to an adequate support being given to the Observatories from public funds, the Directors succeeded in raising sufficient money to keep the two stations at work until the decision of the Committee should be obtained, the subsidy of £350 from the Meteorological Council being also temporarily continued.

The Directors naturally anticipated that the question which led to the institution of this Committee, would be adequately dealt with in the subsequent report. Some of their number, including the two Secretaries, were examined, and fully stated their case, besides handing in detailed memoranda regarding the history, work, and cost of maintenance of the Observatories. After considerable delay, the Findings were issued, but it was no easy matter to reconcile these with the statement of facts lodged in process by the Directors; indeed, the report in semblance was enigmatical, as may readily be judged from the following excerpts therefrom :—

1. "The history of these two Observatories shows that out of the total cost of £24,000, since their establishment in 1883, £17,000 has been received in subscriptions from the surplus fund of the Edinburgh Exhibition, from scientific societies, and from private individuals. During four years, 1888-1902, the Directors were enabled to continue the work of the Observatories only by the generous and liberal assistance of Mr. Mackay Bernard of Dunsinnan. An offer was received from an anonymous benefactor to

defray the entire expense of the Observatories until October, 1904, pending the report of this Committee, in the event of the grant hitherto given by the Meteorological Council being withdrawn, as was proposed, in October, 1903.

2. "We recognise the praiseworthy liberality of those who have contributed so handsomely to the establishment and upkeep of these Observatories, and also the energy displayed by the management and staff in maintaining certain observations for a period of twenty years.

3. "We have most carefully considered the important question whether the Observatories should be retained or abandoned. The evidence on the subject is conflicting. Lord Kelvin still holds to the opinion expressed at the meeting of the British Association in 1887, that 'the Ben Nevis observations are of the highest utility in the development of meteorology, and of framing forecasts of storms and weather,' and he is of opinion that it is a matter for regret that these observations have not been used by the Meteorological Council in preparing their forecasts and warnings. On the other hand, Professor Schuster is of opinion that 'the problems which could with convenience be carried out

at Ben Nevis Observatory have been dealt with,' and that further observations would be superfluous unless some definite problem were set for solution. Both these views received support from witnesses of high scientific standing, and, having regard to the conflict of opinion upon the utility of maintaining the Observatories, we are not surprised that the Meteorological Council, oppressed with financial responsibilities, and desirous of effecting economies, should have come to the conclusion that their annual grant of £350 to the Observatories might be withdrawn without detriment to meteorological science.

4. "But, on the other hand, we find, after weighing all the circumstances, that, on public and economic grounds, it would constitute a bad bargain to allow the Observatories to disappear. Such a proceeding would involve the sacrifice of a large capital outlay which has been expended upon works, and the loss of the property and 'good-will' subsisting in the maintenance of the building and plant. It appears that only £350 per annum is required to insure the continued maintenance of the Observatories, and we are of opinion that every effort should be made to provide this small sum for the purpose.

5. "We are aware that full accord has not always existed in the past between the Meteorological Council and the Directors of the Ben Nevis Observatory, and we are of opinion that, as a condition precedent to a continuance of the grant of £350, arrangements should be made for (1) a re-organisation of the management of these Observatories, (2) a consideration of the general scientific purposes for which they might be used, (3) framing, with the assistance of meteorologists, a scheme defining the lines of investigation which should in future be undertaken at Fort-William and Ben Nevis, and (4) securing to the office the full right to publish telegraphic reports from Ben Nevis along with those from other stations."

The Directors lost no time in calling the attention of the First Lord of the Treasury to the "inexplicably erroneous" statement that "only £350 per annum is required to insure the continued maintenance of the Observatories"; and in appealing to him that means should be found to prevent the abandonment of the Observatories. The Treasury, however, could not see its way to any further increase of the contribution from the Parliamentary grant, but offered to continue the allowance of £350 a year hitherto received. As this arrangement would have left the Directors exactly where they were before, face to face with the impossibility of continuing to raise £650 every year, and with the obvious hopelessness of obtaining anything like sufficient pecuniary support from the Government, there was no alternative but to close the Observatories, which was done on 1st October, 1904.

To not a few it was a matter of profound disappointment that, in a wealthy country such as ours, it should have been found impossible to obtain the comparatively small sum required to carry on a work of great scientific value and interest; but the compulsory disposal of the buildings and dismissal of the staff was, to the Directors, the hardest task of all.

The close bearing of the Ben Nevis observations upon European meteorology was recognised by the German Meteorological Office, which applied for and regularly received by telegraph, daily weather readings from the high and low level stations, and on the close of the Observatories they sympathetically expressed regret that these valuable data were no longer available. Had the Ben Nevis records been continued, and compared, by those engaged in framing the daily weather forecasts, with the telegrams received from other stations, it is not too much to believe that they might have been found to furnish valuable data towards an accurate and scientific forecasting of the weather.

'Twas a sad day for the staff, that bleak first of October—sadder than I could ever hope to indicate in a series of straggling sentences; a sorrow with which strangers might not intermeddle—a bitterness of the heart unspeakably sore. Silent instruments, unrecorded observations—death! one might say, for no obsequies could be less cheerless or sepulchral in aspect than that last scene of all. Strikingly coincidental, too, was the printed precept on the calendar for the day, which announced that "All things cometh to an end," and none the less significant was the entry made in the Log-book,—"This day, by order of the Directors, observations were discontinued at this station."

In a heavy snowstorm, the Observatory was barricaded and locked up on Saturday the 8th of October following, and the superintendent, who had been in residence on the summit during the intervening week, supervising the dismantling of the institution, descended the mountain. The removal of goods and gear had to be accomplished by means of horses, on the backs of which, the contents of the Observatory were taken to Fort-William, and placed in the store at the low level station. Since October, 1883, till the date immediately previously mentioned, there had always been some one in residence at the Ben Nevis Observatory, but on the night which saw the locking up, no human being disturbed the pristine loneliness of the mountain-top.

The feelings and thoughts of the superintendent, as he witnessed the extinguishing of the fires which had glowed for nearly a quarter of a century, and for the last time locked up the doors of the domicile which to him, in common with other members of the staff, had become so dear, may be better imagined than described. Without, the elements were boisterous, and, look where one might, the feeling which seemed always to obtain the mastery, was one of gloom, and the moan and sough of the wind bespoke their own tale of sadness, as if to subscribe their pæan of regret on the occasion. Nearly six inches of snow then lay on the summit, and the wind, which carried the fine powdery flakes in clouds up the precipices and amid the gullies, soon formed wreaths over the bridle-path to a depth of nearly two feet.

The parting of observers and Observatory, moreover, took place on just such another day as that on which the institution was first opened by Mrs. Cameron-Campbell of Monzie; but since then, how many happy hours had been passed in that resort of unique situation? To bright genial days, with balmy airs, long afterglows, and unrivalled sunsets; to the glorious spread of mountains, lakes, and seas; to the rolling fog-banks and serrated ridges tinged with gold from the rising sun—to such charms of retrospect would the observer more willingly let his fancy recur, rather than to those more awesome days when wild Nature held sway. Many a storm had that snugly-built house withstood, many a genial hour had been passed around its cosy ingleside, and many a toil-stained climber had invoked blessings on its inmates and the creature comforts they were on occasion able to bestow.

Greater works than these, however, had been accomplished, and future scientists may yet acknowledge their indebtedness to the records obtained at Ben Nevis Observatory. One can hardly conceive that such an admittedly useful institution is to be abandoned, and left as a haunt of weasels and stoats —cast aside, and allowed to crumble and deteriorate by reason of the rigorous climatic conditions there obtaining. Should such turn out to be the case, the dismantled Observatory will itself be a silent witness to the niggardliness and bungling of State officialism, as well as a slur on the boasted philanthropy of Scotsmen.

VALE.

The material originally positioned here is too large for reproduction in this reissue. A PDF can be downloaded from the web address given on page iv of this book, by clicking on 'Resources Available'.

Printed in the United States
By Bookmasters